Special Concretes
Workability and Mixing

Other books on fresh concrete

Properties of Fresh Concrete
Edited by H. J. Wierig

The shift in concrete production from the site to ready-mixed plants has improved efficiency but has given rise to new problems, such as the influence of temperature and time on consistency of fresh concrete. The separation of mixing and transportation from on-site placing and curing requires better means of measuring and defining the state of the concrete at the time of delivery. In addition, new types of cements and admixtures are being widely used. This book reviews the state-of-the-art of the properties of fresh concrete.

Hardback, 1990, 0-412-37430-7, 400 pages

Rheology of Fresh Cement and Concrete
Edited by P. F. G. Banfill

Understanding and controlling the rheology of cement-based materials is essential if they are to be used successfully in building, civil engineering and offshore applications. This book brings together research information on the flow behaviour of cementitious materials including: measurement techniques and fundamental studies of rheology; hydration, setting kinetics and computer simulation; materials; and practical applications.

Hardback, 1990, 0-419-15360-8, 384 pages

Workability and Quality Control of Concrete
G. H. Tattersall

This book develops the treatment of concrete workability as a property to be measured in terms of two constants. The scientific basis is simply explained and used for the description of practical methods and apparatus for elucidating problems of workability and its use in quality control. The validity and limitations of standard workability tests are fully considered and there are several chapters on the effects on workability of the properties and proportions of mix constituents.

Hardback, 1990, 0-419-14860-4, 272 pages

For further information on these books, contact
The Promotion Department, E & FN Spon, 2–6 Boundary Row, London SE1 8HN, Tel: 071-865 0066.

Special Concretes
Workability and Mixing

Proceedings of the International RILEM Workshop
organized by RILEM Technical Committee TC 145,
Workability of Special Concrete Mixes, in collaboration with
RILEM Technical Committee TC 150, Efficiency of Concrete Mixers,
and held at the University of Paisley, Scotland.

Paisley, Scotland
March 2–3, 1993

EDITED BY

Peter J. M. Bartos

Advanced Concrete Technology Group,
Department of Civil Engineering,
University of Paisley, Paisley, Scotland

E & FN SPON
An Imprint of Chapman & Hall

London · Glasgow · New York · Tokyo · Melbourne · Madras

**Published by E & FN Spon, an imprint of Chapman & Hall,
2–6 Boundary Row, London SE1 8HN, UK**

Chapman & Hall, 2–6 Boundary Row, London SE1 8HN, UK

Blackie Academic & Professional, Wester Cleddens Road, Bishopbriggs,
Glasgow G64 2NZ, UK

Chapman & Hall Inc., One Penn Plaza, 41st Floor, New York NY 10019,
USA

Chapman & Hall Japan, Thomson Publishing Japan, Hirakawacho
Nemoto Building, 6F, 1-7-11 Hirakawa-cho, Chiyoda-ku, Tokyo 102, Japan

Chapman & Hall Australia, Thomas Nelson Australia, 102 Dodds Street,
South Melbourne, Victoria 3205, Australia

Chapman & Hall India, R. Seshadri, 32 Second Main Road, CIT East,
Madras 600 035, India

First edition 1994

© 1994 RILEM

Printed in Great Britain at the University Press, Cambridge

ISBN 0 419 18870 3

A catalogue record for this book is available from the British Library

Library of Congress Cataloging-in-Publication data available

Printed on permanent acid-free text paper, manufactured in accordance
with ANSI/NISO Z 39.48-1992
(Permanence of Paper).

Contents

Preface

Special concretes are a fast growing area of concrete technology where a stage has been reached in which the practical construction industry has to adjust its site practice in order to maintain efficiency and high quality of its products while using a wide range of mixes very different from the traditional ones.

This book consists of papers presented at the International Workshop on Special Concretes: Workability and Mixing. The Workshop was held under the auspices of the International Union of Testing and Research Laboratories for Materials and Structures (RILEM) at the University of Paisley in Paisley, Scotland, on 1–3 March 1993.

The Workshop marked the end of the first year of activity of the RILEM Technical Committee 145-WSM on Workability of Special Concrete Mixes in which it concentrated on the survey of the current knowledge and information. The Workshop brought together invited experts in the field of special concrete who all took an active part in presentation of papers, practical test demonstrations, and an extensive discussion in a true workshop format. The focus of the Workshop was on practical aspects and most of the participants were from industry and had a direct experience with the production or applications of the special concretes. An appropriate fundamental scientific support was also provided by the industrial and academic researchers present at the Workshop.

The organization of this event was a dedicated team effort. The technical programme and the selection of the contributions, followed by the chairmanships of the sessions were carried out by the members of the TC 145-WSM and I wish to thank particularly Dr David Cleland (Belfast), Dr S. Karl (Darmstadt), Dr Claude Legrand (Toulouse) and Mr Örjan Petersson (Stockholm) who devoted much of their personal time and energy and helped to organize the Workshop at very short notice. The cooperation with TC 150-ECM through its Chairman, Professor H. Beitzel (Trier), enabled the mixing process to be included in the theme of the Workshop and made it even more relevant to the needs of practical construction.

Without a principal commercial sponsor, the event relied financially on the goodwill of the University of Paisley which kindly agreed to underwrite the costs of the Workshop. This and the support of the many individual small sponsors who made the event financially viable are gratefully acknowledged.

The complex logistics of the Workshop were managed efficiently by Mrs C. A. MacDonald of the Technology and Business Centre assisted by members of the Advanced Concrete Technology Group and I am also grateful for the practical assistance with the processing of the contributions for publication provided by Mr J. N. Clarke, Senior Editor of E & FN Spon, the publishers.

The response to the Workshop has been very gratifying and the participants' contributions covered the wide range of the special concretes now available. There were more interested potential participants than allowed for by the limit of fifty which had to be imposed to maintain the workshop format of the event. The success of the Workshop ultimately depended on the enthusiasm shown by all of the participants and on behalf of the Organizing Committee and myself I would like to thank everyone who came to the Workshop, contributed their time to prepare and present papers, carried out practical demonstrations and shared their valuable experience and opinions in the lively debates.

Peter J. M. Bartos
Chairman, RILEM TC 145-WSM

Workshop Participants

Members of RILEM Technical Committee TC 145 on Workability of Special Concrete Mixes are noted thus (M)

Dr P. J. M. Bartos (Chairman of RILEM TC 145)
University of Paisley, Paisley, Scotland

Mr D. Beaupré (M)
The University of British Columbia, Vancouver, Canada

Professor H. Beitzel
Fachhochschule Rheinland-Pfalz, Trier, Germany

Mr P. Bennison (M)
Flexcrete Limited, Preston, UK

Mr A. D. R. Brown
Civil & Marine Slag Cement Ltd, Newport, Wales

Mr M. Carlsson
Scancem Chemicals, Oslo, Norway

Dr D. J. Cleland (Secretary of RILEM TC 145)
The Queen's University of Belfast, Belfast, Northern Ireland

Professor W. Cranston
University of Paisley, Paisley, Scotland

Mr A. E. Dearlove
Blue Circle Industries PLC, Greenhithe, UK

Mr A. Garcia
Feder Beton, Montpelier, France

Professor O. E. Gjørv
The Norwegian Institute of Technology, Trondheim, Norway

Mr K. Gunther
Liebherr Mischtechnik GmbH, Bad Schussenried, Germany

Mr M. Hayakawa (M)
Taisei Corporation, Yokohama, Japan

Mr S. Helland
Selmer A/S, Oslo, Norway

Dr D. W. Hobbs
British Cement Association, Crowthorne, UK

Mr L. Hodgkinson (M)
Cormix Construction Chemicals Ltd, Warrington, UK

Mr C. Hua
École Nationale des Ponts et Chaussées, Noisy-le-Grand, France

Mr K. Iversen
Rannsoknastofnun Byggingaridnadarins, Reykjavik, Iceland

Professor C. D. Johnston (M)
The University of Calgary, Calgary, Alberta, Canada

Dr D. Johnston
Pozament Limited, Swadlincote, UK

Mr P. Jones
Castle Cement, Mold, UK

Mr R. Jones
Sika Ltd, Welwyn Garden City, UK

Mr K. Juvas (M)
Partek Concrete, Pargas, Finland

Dr S. Karl (M)
Institut fur Massivbau, Darmstadt, Germany

Professor J. Kasperkiewicz
Polish Academy of Sciences, Warsaw, Poland

Dr F. de Larrard (M)
Laboratoire Central des Ponts et Chaussées, Paris, France

Dr C. Legrand (M)
Laboratoire Materiaux et Durabilité des Constructions, Toulouse, France

Mr P. L. Male
Tarmac–Topmix, Ettingshall, UK

Dr A. McLeish (M)
WS Atkins Structural Engineering, Epsom, UK

Mr G. McWhannel
Fibermesh Europe Ltd, Chesterfield, UK

Mr J. H. Mork
Euroc Research AB, Slite, Sweden

Mr D. Nemegeer (M)
N. V. Bekaert S.A., Zwevegem, Belgium

Mr E. Nicholson
Tilcon Ltd, Glasgow, Scotland

Mr N. H. Nielsen
Skako A/S, Faaborg, Denmark

Dr J. Norberg
Cement och Betong Instituet, Stockholm, Sweden

Mr T. Osterberg
Cementa AB, Vaxjö, Sweden

Mr G. Peiffer
Centre de Recherches de Pont-à-Mousson, Pont-à-Mousson, France

Mr I. D. Peter
Powersprays Ltd, Avonmouth, UK

Mr Ö. Petersson (M)
Cement och Betong Instituet, Stockholm, Sweden

Mr G. Prior
Castle Cement Ltd, Coatbridge, Scotland

Mr M. Rollet
Ciment Lafarge, Saint Cloud, France

Mr S. Smeplass (M)
The University of Trondheim, Trondheim, Norway

Dr L. M. Smith
Scottish Nuclear Ltd, East Kilbride, Scotland

Dr A. K. Tamimi
University of Paisley, Paisley, Scotland

Professor H. Troeber
Fachhochschule Rheinland-Pflaz, Trier, Germany

Dr O. H. Wallevik (M)
Rannsokastofnun Byggingar Idnadarins, Reykjavik, Iceland

Professor J.-D. Wörner (M)
Institut für Massivbau, Darmstadt, Germany

Mr W. Zhu
China Academy of Building Materials, c/o University of Paisley,
Paisley, Scotland

PART ONE
INTRODUCTION

1 WORKABILITY OF SPECIAL FRESH CONCRETES

P. J. M. BARTOS
Advanced Concrete Technology Group, Department of Civil
Engineering, University of Paisley, Paisley, Scotland, UK

Abstract
Significantly modified, 'special' concretes often show unusual
behaviour when in fresh state and the production and quality control
on construction sites cannot rely on tests suitable for ordinary
concrete mixes in ordinary applications. The background to the
increased share of the 'special' concretes in the current construction practice is reviewed, the special concretes defined and the
importance of good knowledge of the production process of special
concretes from batching & mixing to final placing including appropriate and effective testing methods is pointed out.
Keywords: Workability, Testing, Mixing, Quality, High Performance,
Special Concrete, Mix Composition.

1 Introduction

A significant and steadily increasing proportion of concrete used in
construction has been substantially modified to produce mixes with
properties both in their fresh and/or hardened states much different
from those of the ordinary ones. The traditional cement - water -
aggregate mix-design was altered and new ingredients, either singly
or in combinations were introduced into the mix. The modifications
were designed to improve the quality of the concrete which in most
circumstances meant making concrete perform better in more demanding
conditions of service and become more suitable for handling by
specialised methods of construction.

The conditions of service demanded higher properties of the
finished product, namely strength, toughness and durability. The
specialised construction methods demanded mixes suitable for an easy
and reliable placing by pumping, spraying, underwater placing,
extrusion etc.

Such demands and the consequent developments produced a range of
concretes which can be broadly classified as the **'special'** rather
than the **'ordinary'** ones which still prevail in the bulk of everyday
construction.

It is important to appreciate that there are no firm boundaries
between the **special** and **ordinary** concrete mixes. In some situations
one and the same concrete mix can show both an **ordinary and** a **special**
behaviour, depending on which properties are considered important or
in which type of practical application the concrete is used.

Special Concretes: Workability and Mixing. Edited by Peter J. M. Bartos. © RILEM.
Published by E & FN Spon, 2–6 Boundary Row, London SE1 8HN, 0 419 18870 3.

2 Definition of a special fresh concrete

For practical purposes of setting out a rational programme of appraisal and investigation an appropriate separation of the **special** mixes from the **ordinary** ones has to be established. In the context of this book, and in accordance with a previous decision of the RILEM Technical Committee TC 145-WSM (Bartos 1993) the special fresh concretes were defined as concretes which in their fresh state cannot be adequately assessed by one or more of the common standardised workability tests. The common tests include the **slump test,** the **VeBe test** and the **Spread/Flow-table test**.

The above definition indicated the following **special concretes**:
>High strength concrete (Compressive strength grade > 80 MPa)
>Non-dispersible underwater concrete
>Foamed concrete
>Flowing, superplasticized concrete
>Sprayed concrete
>Cementitious concrete repair mixes
>Very dry mixes for precast concrete
>Roller compacted concrete

Concurrently, concretes which contained significant proportions of **special ingredients** have been also identified as **special concretes**. According to their composition, fresh concretes containing the following special constituents were included:

>Fibres of all types (steel, glass, polymer etc.)
>Microsilica (silica-fume)
>Polymers
>Special cements
>Very low or high density aggregate
>Fine fillers
>Fly ash and slag (in high proportions)

3 Workability of special concretes and the construction process

There is no shortage of information on properties of **hardened special concretes,** such as the high strength, fibre reinforced and other mixes generally described as 'high perfdormance' concretes. Numerous publications are available and there is a substantial amount of reliable evidence proving that the range of outstanding properties of the special hardened concrete can be obtained not only in the laboratory but also in a real construction.

However, information on workability of the special concretes tends to be scattered in the very many publications on hardened concrete. Such sources generally mention the production method for a particular type of concrete but any workability parameters, if noted at all, are chosen arbitrarily and would have been very rarely investigated.

Books specialising in workability tended to focus more on the fundamental aspects of rheology of ordinary mixes rather than on practical workability parameters necessary for the construction using the special mixes. A recent book (Bartos 1992) has concentrated on workability and practical tests for fresh concrete in general and the author has also surveyed a number of practical tests applicable to several of the fresh special mixes.

An important source of information about the special concretes are the product data sheets provided by suppliers of ingredients which turn the ordinary mixes into the special ones. Such information focuses mainly on the properties of the final product and the actual concrete production process is neither normally covered in sufficient detail to assist in its practical execution nor, for commercial reasons, does it offer or discusses alternatives.

All the conclusions point to a clear lack of independent, practical guidance for contractors, engineers and other users regarding the specification of properties of fresh special concretes, their significance for the construction process and the means for their control and verification.

The performance of the special mixes in their fresh state is invariably critical for a successful execution of the construction and a satisfactory long-term performance of the material once it has hardened.

The behaviour of a special mix when fresh and its final properties are in many cases strongly influenced by the **mixing process** which has to cope with the distribution of small quantities of additional ingredients uniformly throughout the mix. The same situation as described above for the workability of the fresh mixes applies to the case of mixing, although there is much less information available. The types of mixers, batching sequences and the speed of mixing are some of the main para- meters about which there is little practical guidance. In collaboration with the RILEM TC 150 ECM on Efficiency of Concrete Mixers a common aim of developing practical guides for **production of special concretes** has been adopted. It will cover the whole process, from batching and mixing to handling of the fresh mix and the relevant tests for production parameters and quality control.

The case of a **non-dispersive underwater concrete** illustrates well the main points about the significance of workability. There are now several specialist admixtures available freely on the construction market which convert a suitable ordinary mix into the special, washout-resistant, non-dispersive mix and hundreds of thousands of m3 of such special concrete are placed in a year. Practical experience has confirmed the high degree of site control required and demonstrated the high cost of remedying failures of underwater concretes. It has also provided a clear indication that none of the current standard workability tests can effectively be used to assess meaningfully and reliably the fresh non-dispersive underwater concrete.

Despite the difficulties encountered, no alternative standard test exists and no test has been even generally accepted. Moreover, there is no standard or a generally accepted national test for the assessment of the resistance of this mix to washout when placed under water. Each R&D into the special 'underwater' admixture used a different test and it is therefore impossible to confirm independently and compare the performance of the special admixtures and mix compositions available.

4 Conclusions

This book represents one of the early stages in the process of collection and appraisal of the existing information, identification and subsequent filling of any large gaps in the knowledge of the mixing process and properties of fresh special mixes undertaken by the RILEM TC 145 WSM and 150 ECM Intl. Technical Committees.

It also provides a significant review of the current practice through contributions by many experienced practitioners with an additional support from researchers working on practical problems related to development and applications of the special fresh concrete mixes.

The book deals with most of the **special mixes** previously identified, the coverage extends from mixing and batching to handling and testing of the special fresh concretes.

5 References

Bartos, P.J.M. (1993) Workability of special concrete mixes, **Materials and Structures**, 26, 50-52.

Bartos, P.J.M. (1992) **Fresh Concrete: Properties and tests**, Elsevier Science Publishers, Amsterdam, The Netherlands.

2 AUTOMATION OF THE CONCRETE CONSTRUCTION SITE

H. BEITZEL
Institut für Bauverfahrenstechnik, Fachhochschule Rheinland
Pfalz, Trier, Germany

Abstract
Automation of the concrete construction site has clearly
increased over the past eight years. The economic course
of the concrete's preparation and installation is determined by new technology in the field of concrete mixing
plants, truck-mixer steering systems and concrete pumping
systems. In the future the computer will be of indispensable
assistance to the site engineer.

1 Introduction

The economic success of a concrete building site is determined above all, by the rationalisation and automation
of the stages involved in the processes of preparing,
transporting and utilizing the fresh concrete.

A dinstinction is made between ready-mixed concrete
and concrete mixed in transit. The former can be manufactured not only at the central production plant but
also at the concrete mixing plant on the batch construction site. With the latter, concrete mixed in transit,
the batch components are fed via a dosage apparatus into
the truck mixer; at the construction site water is added
and the batch materials are checked, mixed and discharged.

2 Ready-mixed concrete – concrete mixed in transit

Ready-mixed concrete is chiefly manufactured in automated
batching and mixing plants. Nowadays, the nature of the
plants commenly employed, als well as the large number
of fixed mix formulae require the intendified use of computers to regulate the manufacturing process. Only the
engineering process, such as the dosage, weighting, mixing
and emptying are still superintended by the operator
(Fig.1.).

The moisture content of the aggregates and the consistency of the concrete are automatically measured by
the manufacturers. However, the methods used vary from
one manufacturer to another and unfortunately, the results

Special Concretes: Workability and Mixing. Edited by Peter J. M. Bartos. © RILEM.
Published by E & FN Spon, 2–6 Boundary Row, London SE1 8HN, 0 419 18870 3.

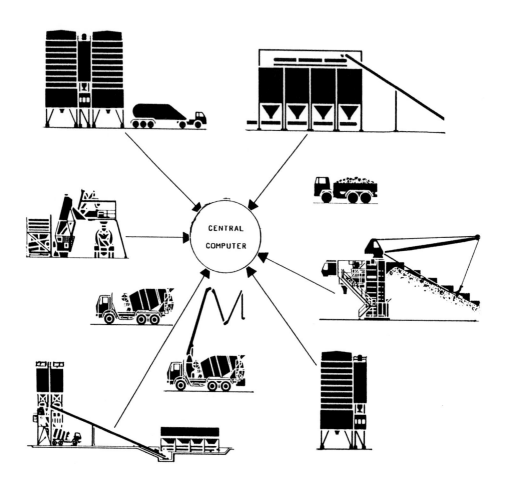

Fig. 1.   Computers to regulate the manufacturing process

of these measurements are to some extent not accurate enough.

A high working cycle rate and good performance ensure the optimum utilization of the concrete mixer. When planing, it is this utilization which is placed in the foreground. Despite the major importance and the seemingly high level of technical development attributed to such machines, their use is, to some extent, still burdened by uncertainities regarding quality (Fig.2.).

More recent techniques for testing and measuring assist in supplying information about the various qualities of concrete, whereby the manufacture according to different mixing methods and the mixing times were taken into account.

New developments focus on improving the disposability and operational safety of the machines, the saving of energy and changes in process engineering. Other topics under current research are the service life of wearing parts when using substitute materials instead of aggregates and the measurement of the moisture content.

In many countries the batch components are put as dry materials into the mixing drum of the truck mixer; the water is filled into a separate tank and the two are mixed at the construction site. Especially in torrid zones, transport over lang distances or in temperate zones where the times calculated for transport cannot always be adhered to, it is neccessary to use ready-mixed concrete to avoid unwanted setting of fresh concrete during transport.

The issue as to wether one should give preference to ready-mixed concrete or concrete mixed in transit -when taking into consideration the quality of the concrete and the investment and running costs- has repeatedly ended without a solution. Nowadays, there are truck-mixers available which are controlled and supervised by computers using improved process engineering techniques, Fig.3.

The truck-mixer was specially developed for application in the Twin-Fin Process Mixer GSM with counter-current mixing system. It helps to produce high-quality concrete and guarantees precise water batching and compliance with present values, such as desired water value, mixing time and number of drum revolutions. All important data are stored and logged by a paper tape printer.

3 Mobile concrete pumps

The requirements of the concrete pumps regarding the effeciency of depositing the concrete, the distance and heigh to which the concrete has to be conveyed have greatly increased. These requirements refer to all branches of civil engineering, tunnelling and the building of power station (Fig.4.).

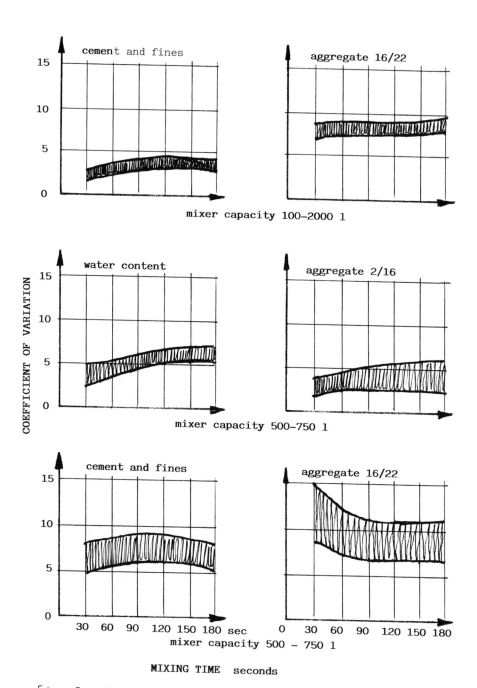

Fig. 2. Results:Pan - and Compulsory Mixers

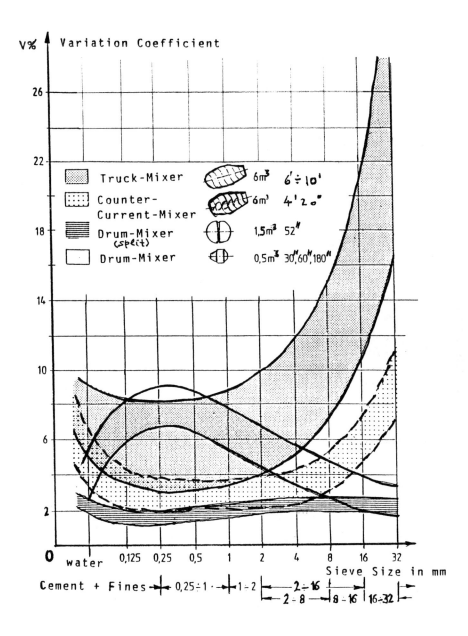

Fig. 3. Results Truck-Mixers and Drum-Mixers

The types of concrete pumps available have no problems with operational safety when normal concrete formulae are applied. Yet, a few concrete pumps show difficulties in conveying, when special types of concrete ar used.

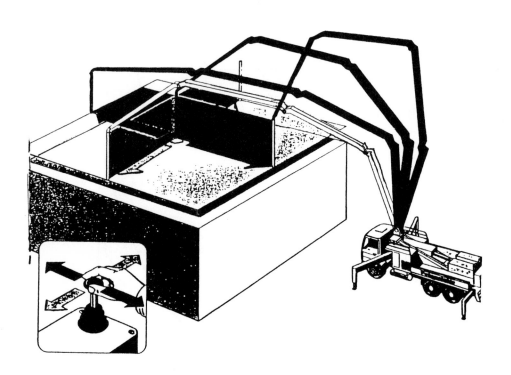

Fig. 4. Concrete Pumps

At present, mobile and stationary distribution masts which are driven by a three-coordinate-steering-system can be used to reach all working areas. Choosing the so-called TIT (Teach-In-Technique) it is also possible to reach working areas fully automatically.

In the future, the site engineer will not be able to dispense with the computer at the building site, and thus one can almost expect automated concrete construction sites, at least in certain sectors.

12

4 Summary

Automation of the concrete construction site has clearly increased over the past eight years. The economic course of the concrete's preparation and installation is determined by new technology in the field of concrete mixing plants, truck-mixer steering systems and concrete pumping systems. In the future the computer will be of indispensable assistance to the site engineer.

5 References

Beitzel, H. Einsatz von Betonmischer, Schweizer Ingenieur und Architekt, Heft 41

Beitzel, H. Systemspezifische Bewertung bei Betonpumpen, Straßen- und Tiefbau, Heft 7/8, 1986

PART TWO

MIXING PROCEDURE AND EVALUATION OF MIXERS

3 MEASUREMENT OF PROPERTIES OF FRESH HIGH PERFORMANCE CONCRETE AND EFFECTS OF THE MIXING PROCESS

T. ÖSTERBERG
Cementa AB, FoU Betong, Danderyd, Sweden

Abstract
In order to determine how silica fumes - compacted, non-compacted and in water suspension (slurry) - affected the consistence of a high performance concrete mix, a series of approximately twenty mixtures were prepared under laboratory conditions. The series included two cements of low alkali type and with low contents of C_3A. The investigation also included three different mixing successions. The properties of the fresh concrete were measured by a slump and with a BML-type series 2 concrete viscometer, over a time domain of 5 to 60 minutes. The results indicate that silica slurry yields the most favourable consistencies if the properties of the concrete in fresh state are characterised in terms of slump, yield value and viscosity. The mixing procedure giving cement the best possible prospect to dispere in the paste phase, proved to be the most favourable. The choice of cement affected the consistencies in a conspicuous way. The more finely ground P 400 resulted in a significantly higher yield value in the mixtures and lower slump values.

1 Introduction

In order to determine how silica fume in different states (compacted, non-compacted and in water suspension [slurry]), affects the consistence in fresh state of a high performance concrete, a number of mixtures were prepared under laboratory conditions. The properties of these mixtures were measured by slump and with a BML-type series 2 concrete viscometer in its original configuration.

The project included two different cements and the testing of three different mixing sequences. The purpose of the latter was to vary the dispersion of cement and microsilica.

Special Concretes: Workability and Mixing. Edited by Peter J. M. Bartos. © RILEM.
Published by E & FN Spon, 2–6 Boundary Row, London SE1 8HN, 0 419 18870 3.

The investigation plan has been established in co-operation with the Swedish Cement and Concrete Research Institute (CBI), and is a part of the Swedish national research project "High Performance Concrete", sub project 1; "Particle Packing and Rheology".

2   Concrete composition

The concrete composition is based on reference mix proportions given in the Swedish national research project "High Performance Concrete":

Cement		450,0	kg/m^3
Silica (dry)		22,5	"-
Super plasticiser 92 M		13,0	"-
Gravel	0-8 mm	933,0	"-
Crushed granite	12-16 mm	862,0	"-

The gravel comes from a glacier deposit. The petrography is mainly granitic to gneissic, and with minor constituencies of leptite. The crushed material consists of a fine to moderately grained granite. The super plasticiser is a melamine formaldehyde sulfonated condensate of Cementa origin. The solid content is 35%.

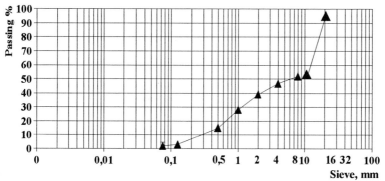

Fig. 1. Compounded aggregate grading.

3   Variables

The project included the following variables:
- Type of cement
- Mixing order (dispersion of the dry components)
- Degree of dispersion of silica particles

These variables are described in detail below.

3.1 Mixing order

i) Cement and aggregate is dry mixed for 60 seconds. The Mixer is stopped. After this, water and silica (slurry) is simultaneously added . The mixer is started and then runs for approx. 15-20 sec., after which the super plasticiser is added.*

ii) The crushed granite, 50% of the water and silica (slurry) are mixed for a minute. The mixer is stopped. Cement is added and after this the rest of water. The mixing then goes on for 15-20 sec, after which the super plasticiser is added. The mixer is stopped and the gravel is added.*

iii) Cement and crushed granite is dry mixed for approx. 10 seconds. The mixer is stopped. 80% of the water is added and the mixing procedure continues for one minute. Silica (slurry) is added. The mixing then goes on for 15-20 sec., after which the super plasticiser is added.The mixer is stopped. The rest of the water and the gravel is added.*
* Final mixing goes on for approx. 3 minutes.

The mixing was done in a 150 litres Eirich-mixer. Batch volume: 90 litres.

3.2 Silica fumes

Compacted and non-compacted silica (sacked) was acquired from Scancem Chemicals in Oslo. The SiO_2 content was typically > 92%. The slurry was prepared at laboratory by dispersing uncompacted silica in water.

3.3 Cement

The sacked cement was taken from a cool, sheltered storage in Slite. The cement was less than 3 months old.

Table 1

Bogue analysis		Degerhamn Std P	Degerhamn Std P (400)
Na_2O (eqv.)	% of weight	< 0,6	< 0,6
CaO	"	< 1	< 1
C_3S	"	52 - 58	52 - 58
C_2S	"	20 - 24	20 - 24
C_3A	"	1,5 - 1,9	1,5 - 1,9
C_4AF	"	14 ± 1	14 ± 1
$CaSo_4$	"	2,1	2,1
Water demand	%	26	27
Blaine	m^2/kg	300	405
1-day strength	MPa	10	16
28-day strength	MPa	55	61

4   Testing program

4.1 Concrete properties in fresh state

a) The slump was measured on concrete 1, 15, 30 and 60
   minutes after concluded mixing. The concrete was
   stored in the mixer (with the lid on) in between
   tests. Prior to each test, the mixer was run for
   approx. 5 seconds.
b) The fresh concrete was also subjected to a concrete
   viscometrical examination with determination of g- and
   h-values. This test was carried out with the same time
   intervals as those of (a). Storage and mixing prior to
   testing according to (a).
c) A subjective view of the consistence was noted at the
   conclusion of the mixing

4.2 Casting of test specimens

For each batch, 15 specimens includes 150 mm cubes and 2
beams for determination of the flexural strength were cast
in connection with the first measurement of consistence.
The moulds were filled in two layers. The compaction took
place on a vibration table. The moulds were subjected to
vibration approx. 30 sec. per layer, or until large air
bubbles had ceased to surface. The test specimens were
stored according to Swedish Standard, SS 13 72 10 (+20°C
dry storage conditions), until the testing occurred.

4.3 Strength testing

Strength testing was conducted according to SS 13 72 10 or
the applicable Swedish standard. Cube strength was
determined at an age of 1, 3, 7, 28 and 91 days
respectively. The flexural strength was determined at an
age of 28 days.

5   Results

In table 1 and 2 and in the supplements of this report, the
results of slump measurements, the concrete viscometry and
strength tests are described.

5.1 Slump measurements

Graphical presentations of consistence progress, measured
in accordance to slump are shown in figures 2, 3 and 4. In
the graphs depicted, slump has been set aside as a function
of the mixing order and the state of silica fume. The
binder used in the first three figures is Standard P
Degerhamn.

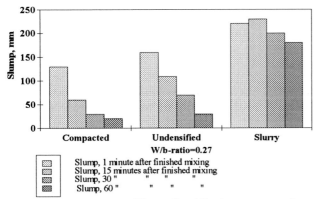

Fig. 2. Mixing procedure **i**.

Figure 2 displays a marked trend indicating that silica in water suspension (slurry) yields the most favourable initial consistence and also the least change of slump over a time period. This follows by non compacted silica fume. A feasible explanation to this is that the mixing procedure (i), results in a high dispersion of cement, but a lesser dispersion of the silica fume.

Using mixtures including compacted silica may result in non-dissolved agglomerate and lumps of silica particles which in turn may affect the result in a negative way. The mixture containing silica slurry, incorporates well dispersed silica particles (separate dispersion, prior to mixing) from the start and the cement is well dispersed, owing to the mixing procedure.

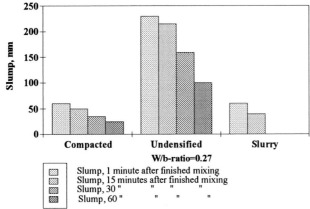

Fig. 3. Mixing procedure **ii**.

The result depicted in Fig. 3 is more difficult to interpret. Here the non-compacted silica fume has rendered the most favourable result. The reason why the silica in the slurry state did not yield the result expected may seem peculiar.

There was in this case reason to suspect a test related artefact. The mixing was later repeated and the result showed to be more or less identical with that of the first. Evidently, the result was to be regarded as invariant. The explanation of this remarkable result may possibly be found in the following: in the case in which silica slurry has been used, the friction of the particles were too low during the early stages of the mixing, due to the large quantity of liquid.

This may have resulted in a creation of lumps of cement when this was added to the mixer. Presumably, the lumps were never dispersed during the mixing procedure and contributed to a negative influence on the consistence (higher inner friction). A slightly lower strength after 28-days of maturing supports this assumption (see Table 2).

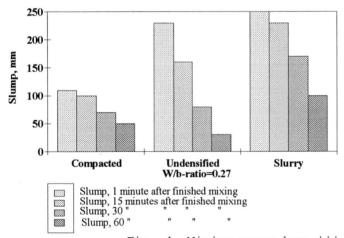

Fig. 4. Mixing procedure **iii**.

The mixing procedure **iii**), has yielded the most favourable results, overall. The mixing procedure incorporates good possibilities for the cement particles to disperse during the dry mixing and the initially high amount of water gives good possibilities to wet a majority of the particles. The crushed aggregate prevents agglomeration when the silica is added. Again, the concrete with compacted silica fume shows the least favourable consistence, probably due to silica lumps.

Figures 5, 6 and 7 depict a corresponding slump measurement of the concrete with P 400. It is evident that

a change of binder, from Degerhamn std P to P 400, has
generally resulted in a obviously negative influence on the
initial consistence.
The only batches showing reasonably comparable results to
the batches using Standard P Degerhamn, were those prepared
using mixing procedure **iii**.

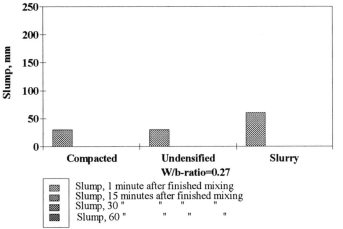

Fig. 5. Mixing procedure 1 (**ii**).

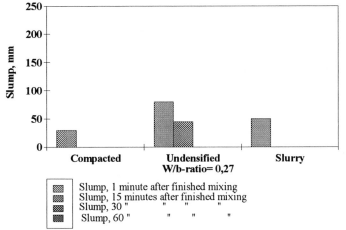

Fig. 6. Mixing procedure 2 (**ii**).

23

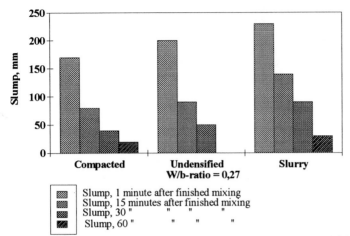

Fig. 7. Mixing procedure 3 (**iii**).

The explanation to this is related to the water demand of the cement. Assuming a favourable composition of the aggregate grading, the water demand of the cement and the ability of the system to disperse the finer particles, are the most important factors determining the possibility to obtain a favourable consistence.

This has earlier been demonstrated, e.g. by Fagerlund (1986).

The slump measurements show that a change in consistence also takes place considerably faster in concrete with P 400. The reason to this is likely due to a notably larger reactive surface of the binder. The hydration processes will therefore be initiated more rapidly and more water will be consumed during the early stages of the hydration and forming of reaction products.

5.2 Concrete viscosity measurements

The outturn of these measurements are depicted in tables 2, 3 and 4. Concrete viscosity measurements have been conducted with a concrete viscometer type BML, series 2, built at NTH in Trondheim, Norway. The inner coaxial measuring device used is of original design. The size is 200 * Ø 200 mm. Overall diameter is approx. 400 mm.

Measurement and calculation of viscosity (h-value)

The measurement of viscosity assumes that fresh concrete behaves in a great part, as a Bingham fluid. The measuring procedure and theory of Bingham fluids will not be discussed here (Tattersall [1991]). The h-value is defined as the incline of a line average, obtained by a regression analysis of a number of torque measurements, gathered at separate angular velocities. The measuring procedure and the subsequent evaluation of measure data was described in detail by Wallevik (1990).

The evaluation of data from this segment, is based on measurements of approx. 10 batches. Remaining measurements produced results of a non-reliable nature. These will be further commented on below.

Batch 901 shows a high viscosity at the time of the first measurement (1 min). The viscosity later decreases, as the g-value simultaneously sharply increases. This does not seem probable. During the measurement in question, some sort of coagulation is likely to have occurred in the concrete, yielding high values during the first phase of the measuring process (the highest velocities). This binding has later been released. Alternatively, the concrete has been separated which has resulted in a considerably lower torsion resistance at a lower angular velocity.

Also the results of the measurements of batches 902 and 903 seem peculiar. Viscosity tends to decrease after a time span of 30 and 60 minutes respectively, in place of the expected increase . This too may be explained by the separation and/or the development of slipping surfaces (the forming of plugs). The same is valid in the case of 904.

The results of the measurements and calculations of the viscosity in batches with cement P 400 show the same tendency towards a decrease in viscosity in a majority of the mixes. This circumstance reduces the reliability as to the significance of these measurements. In some cases, plugs were presumably formed as early as during the first measuring sequence. In other cases this occured after 15 or 30 minutes. The measurements of batch 918 show progress with a gradually increasing viscosity. These results appear to be more consistent.

Conclusively, the risk of non-desirable effects exemplified by a separation and the forming of a plug attached to the inner coaxial cylinder is evidently larger when conducting concrete viscosimetry on concrete with low w/b-ratio. These undesirable processes are more seldom seen in conventional concrete with w/b-ratio of approx. 0.45 and greater (Wallwik, 1990).

A tabulation of the viscosity's (expressed as h-values) relation to mixing order and silica state is shown in Table 4 below. The table only includes the measured data from the first measurement (1 minute). Data of inferior qualities referred to above which may have affected the outcome of the measuring, has been screened out.

Table 4 Measured viscosities (*Italics = P 400*)

Mixing procedure	Compacted silica h-value/batch #	Non-compacted silica h-value/batch #	Silica slurry h-value/batch #
i)	53/901	25/904,	18/907
ii)	25/902	25/905,	29/908
		22/914	
iii)	19/903,	22/906,	19/909,
	22/912	*21/915*	*21/918*

From this, admittedly a rather scarce foundation, the following trends may be identified:
- Generally, the mixing procedure **iii)** seems to have yielded the lowest viscosity
- Silica slurry has produced the lowest h - values
- The mixing procedure **i)** including silica slurry has yielded a low viscosity

Calculated g-values

The g-value is a fictitious entity, defined as the point of intersection on the y-axis of the line obtained from a regression analysis as indicated above. Several researchers e.g. Tattersall (1991), Banfill (1990), Murata and Kikukawa (1979) have established that the g-value reflects a material property of the fresh concrete. This property is referred to as the yield value. A relation with the slump has been observed by Murata and others. A plotting of the obtained g-values from batches not suffering from undesirable interferences (batches 905, 907, 910, 915 918), indicates the existence of a relation between the g-value and the slump. Figure 8 depicts g-values plotted against the slump in the same time frame.

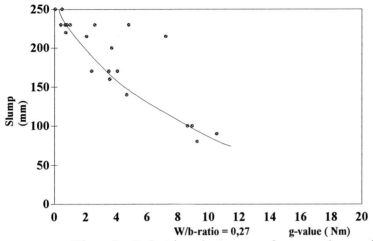

Fig. 8. Relation between slump and g-value

Figure 8 indicates the existence of a certain relation
between slump and g-values, despite the rather extensive
scattering of the data, see also Fig. 7.

A tabulation of obtained g-values in different mixing
procedures and silica state renders the following result:

Table 5 Measured yield values(*Italics = P 400 Cement*)

Mixing procedure	Compacted silica g-value/batch #	Non-compacted silica g-value/batch #	Silica slurry g-value/batch #
i)	12,4/*910*	4,78/904	0,74/907
ii)	9,72/902	2,5/905, 10,0/*914*	12,40/908
iii)	7,33/903	0,03/906, 4,05*1/915*	0,06/909 0,68*1/918*

Values indicating separation and forming of plugs have been
screened out.

From the table data, it is deducible that the mixing
procedure **iii)**, with silica slurry and non-compacted silica
will give the lowest viscosities.
Furthermore, mixing procedure **i)** and silica slurry also
appears to give a rather low values.

6 Conclusions

The result from approx. 20 batches of high performance concrete indicates a clear pattern regarding the properties in fresh state;

- An improved dispersion (distribution) of the finest particles will produce a more favourable consistence.

- A mixing procedure in which the included particles are efficiently moistened, particularly cement, is also favourable.

- The mixing order when adding silica slurry to high performance concrete, is of importance in order to avoid the formation of lumps (coagulation). Silica slurry should not be added prior to the adding of the main bulk of the water.

- Water demand and water consumption of the cement during the initial hydration phase, have great impact on the consistence and consistence progress of the concrete with low w/b-ratio. This is strongly related to the Blaine and the distribution of the different cement minerals (mainly C_3A and C_4AF).

- It possible to produce a high performance concrete with properties in fresh state similar to those of conventional concrete. This assumes a choice of a cement with properties indicated above, silica in predispersed state and an carefully chosen mixing order.

7 References

Fagerlund G. CMT-report 85020. Cementa Danderyd   1985.

Wallevik O. H. and Gjörv O. E. "Modification of   The Two - Point Workability Apparatus". Seminar on   Rheology of Fresh Concrete. NTH Trondheim, Oct. 1990.

Tattersall G. H. Workability and Quality Control of Concrete. 262 p. E & FN Spon publ., London 1991

Banfill P.F.G. Experimental Investigations of   The Rheology of Fresh   Mortar. Proceedings of the   Rilem Colloquium; "Properties of Fresh Concrete". Hanover 1990.

Murata J., Kikukawa K., "Measuring Rheological   Constants of Fresh Concrete Proc.". JSCE, no 284 p.p. 117-126. Tokyo 1979.

8   Appendix A. High performance concrete test results

The test results are described in tables 6 and 7.

Table 6   G- and H-values

901	g	h
1 min.	2.90	53.48
15 min.	28.15	8.70
30 min.	26.39	11.57

902	g	h
1 min.	9.72	25.40
15 min.	16.51	24.26
30 min.	24.93	12.11

903	g	h
1 min.	7.33	18.80
15 min.	9.49	20.63
30 min.	13.04	19.52
60 min.	18.00	15.87

904	g	h
1 min.	4.78	24.61
15 min.	7.21	28.07
30 min.	18.04	21.08

905	g	h
1 min.	2.58	24.51
15 min.	2.05	29.61
30 min.	3.57	36.88
60 min.	8.93	32.01

906	g	h
1 min.	0.03	21.95
15 min.	3.42	28.91
30 min.	7.95	27.99
60 min.	15.30	22.45

907	g	h
1 min.	0.74	17.65
15 min.	0.42	32.40
30 min.	3.70	32.54

908	g	h
1 min.	12.40	29.13

909	g	h
1 min.	0.06	19.13
15 min.	0.80	24.72
30 min.	2.39	29.04
60 min.	8.63	27.42

910	g	h
1 min.	19.04	12.56

912	g	h
1 min.	4.05	21.85
15 min.	9.26	21.94

914	g	h
1 min.	10.03	21.92

915	g	h
1 min.	4.05	20.92
15 min.	10.55	20.03

918	g	h
1 min.	0.68	20.79
15 min.	4.66	23.16
30 min.	10.18	23.01

919	g	h
1 min.	10.00	24.08

Table 7

High Performance Concrete

Batch no.	Type of Cement	Type of Silica	Mixing type	Air content %	Density kg/m³	Slump 1 min. mm	Slump 15 min. mm	Slump 30 min. mm	Slump 60 min. mm	Strength 1 day MPa	Strength 3 days MPa	Strength 7 days MPa	Strength 28 days MPa	Strength 91 days MPa	Flexural strength 28 days 2 pcs. MPa
901	A-Cem	Densi.	1	1.8	2460	130	60	30	20	28.3	66.0	87.3	112.1	120.8	10.3-10.2
902	A-Cem	Densi.	2	1.9	2460	60	50	35	25	27.7	58.0	81.2	107.7	114.7	9.4-10.4
903	A-Cem	Densi.	3	1.7	2450	110	100	70	50	31.4	61.5	81.6	113.3	116.0	10.8-10.5
904	A-Cem	Undensi.	1	1.5	2460	160	110	70	30	27.8	59.9	77.9	108.1	116.2	8.1-9.4
905	A-Cem	Undensi.	2	1.5	2460	230	215	160	100	26.9	59.4	85.2	113.8	118.7	10.0-10.2
906	A-Cem	Undensi.	3	1.5	2460	230	160	80	30	27.0	60.2	81.6	112.0		10.5-10.9
907	A-Cem	Slurry	1	0.8	2480	220	230	200	180	2.8	56.3	83.7	110.4	120.9	10.0-8.8
908	A-Cem	Slurry	2	2.9	2400	60	40	-	-	31.0	59.8	83.6	105.3		10.0-10.8
909	A-Cem	Slurry	3	0.7	2470	250	230	170	100	26.2	60.8	84.0	113.7		9.8-10.5
910	P-400	Densi.	1	2.4	2440	30	-	-	-	40.4	64.6	81.1	108.8		11.0-10.5
911	P-400	Densi.	2	2.3	2430	30	-	-	-	40.9	64.1	83.1	106.7		9.9-10.6
912	P-400	Densi.	3	1.8	2440	170	80	40	20	39.8	66.0	87.3	110.7		8.9-10.6
913	P-400	Undensi.	1	2.0	2450	30	-	-	-	39.4	65.6	89.0	112.6		10.0-11.2
914	P-400	Undensi.	2	1.8	2440	80	45	-	-	37.9	63.8	88.2	110.8		11.0-10.1
915	P-400	Undensi.	3	1.4	2460	200	90	50	-	37.2	66.3	85.8	115.3		10.0-11.1
916	P-400	Slurry	1	2.0	2450	60	-	-	-	39.6	65.4	89.8	113.0		10.7-10.9
917	P-400	Slurry	2	1.9	2430	50	-	-	-	40.5	67.0	91.1	101.2		9.6-9.7
918	P-400	Slurry	3	1.1	2490	230	140	90	30	40.6	68.2	92.9	114.5		9.7-10.0
919	A-Cem	Slurry	2	2.0	2440	50	-	-	-	30.0	57.1				

4 EFFECTS OF TWO-STAGE MIXING TECHNIQUE ON STRENGTH AND MICRO-HARDNESS OF CONCRETE OF CONCRETE

A. K. TAMIMI
Department of Civil Engineering, University of Paisley, Paisley, Scotland, UK
P. RIDGWAY
Department of Civil Engineering, University of Strathclyde, Glasgow, Scotland, UK

Abstract
This research programme is based on a (new) mixing technique in which concrete is produced by mixing water at two separate times. Two-stage concrete can be obtained by premixing cement, sand and coarse aggregate. This concrete compared with the conventional concrete exhibits low bleeding and high rate of strength development in the entire range of mixes studied, but significant improvement in compressive strength is only shown by mixes of high workability. Micro-hardness studies showed that the two-stage mixing technique increased the minimum and maximum micro-hardness of the interface compared with the conventional concrete, which explains the effects of the two-stage mixing technique.
Keywords : Two-stage Mixing, Concrete, Microhardness, Bleeding, Compressive Strength, Workability, Bond, Cement paste Aggregate Interface, Interfacial Zone.

1 Introduction

A new concrete mixing technique was first reported by Higuchi(1980), and then presented at the international conference on "Bond in concrete", held at Paisley College of Technology Scotland in June1982, by Hayakawa and Itoh(1982). The procedure is basically the same as for mixing of conventional concrete, except that the water is divided into two portions, and poured into the mixer at two separate times. The amount of the first pour is controlled to be 25% by weight of the cement to be used including the surface moisture of the sand and gravel. Mixing was performed in a "Cumflow" pan mixer of 125kg capacity. The concrete produced by this method is also called SEC concrete (sand enveloped with cement).

Special Concretes: Workability and Mixing. Edited by Peter J. M. Bartos. © RILEM.
Published by E & FN Spon, 2–6 Boundary Row, London SE1 8HN, 0 419 18870 3.

The primary objective of this investigations was to evaluate the effectiveness of the new mixing method in reducing the bleeding capacity and increasing strength at various ages. All the parameters and variances were kept constant for each comparable mix of the 'two-stage' SEC concrete and conventional concrete.

Properties measured included the compressive, indirect tensile strength, bleeding capacity, workability and micro-hardness tests at 3,7,14 and 28 days with water/cement ratio of 0.45, 0.55 and 0.70. For every case four mixes were prepared, two by the two-stage mixing technique and two by the conventional mixing technique.

For each mixing technique artificial glass marbles were substituted for the 13.0 mm aggregate in one mix, while the second mix remained un substituted in order to obtain standardized identical aggregate size and interface for both mixing techniques to compare them using the micro-hardness measurements.

2 The microhardness technique and preparation of samples

It has been shown that the interfacial bond between the paste and aggregate is an important parameter in determining the properties of a concrete (Carles-Gibergues 1982). The region between the aggregate and the bulk paste has been termed the aureole of transition (Maso 1967).

Micro-hardness measurements were made across the aureole of transition of the glass marble-cement paste for both mixing technique. A Vickers Micro-Hardness tester was used with a load capacity ranging from 0.15 to 20 N. For each mix and after the indirect tensile test, two slices (approximately 20mm thick) were sawn at right angles to the direction as cast from the middle third of the split cylinder using a diamond saw and solution of soluble oil as cooler.

Polishing was accomplished using four horizontal grinding wheels (with abrasive grits No. 60, 300, 600 and 1200). Indentations with a pyramid-shaped diamond were made across the glass marble-cement paste interface in lines radiating from the centre of the glass marble toward the bulk paste in order to obtain at least three readings at each distance from the interface. Using the optical microscope, two reading were taken, the distance from the interface i.e from the centre of the indentation to the interface and the mean value of each indentation diagonals from which the micro-hardness measurements can be obtained using standard hardness charts.

All the diagrams representing results of this investigation were drawn using plotlib computer program of curve best fitting and represent the variation of the micro-hardness measurements with distance from the interface.

Table 1. Results of the Mechanical Tests

Mix Proportions (by mass)	W/C Ratio	Workability		Bleeding Rate (%)	Improvement in Compressive Strength (%)					Change of Tensile Strength 28day (%)
		Slump mm	Ve–Be sec		1day	3day	7day	14day	28day	
1:2:4	0.45	60*	12.5	1.57	5.7	6.4	2.9	7.2	5.4	21.6
		100	10.5	0.66						
1:2:4	0.55	80	8.00	1.84	19	17.5	10.4	8.0	0.0	–3.5
		120	6.00	0.89						
1:2:4	0.70	shear	0.00	3.28	25	21.5	22.5	15.6	5.8	4.5
		collapse	0.00	1.03						
1:3:5	0.45	0.00	20.0	1.07	7.0	1.0	8.3	0.6	5.3	4.3
		0.00	12.0	0.42						
1:3:5	0.55	30	15.0	1.50	13.6	10.4	8.1	5.1	6.9	7.7
		50	11.0	0.61						
1:3:5	0.70	85	7.00	1.87	24.6	23	16.1	14.7	7.2	10.0
		120	3.50	0.77						

* Upper values for the conventional mix & the lower values for the two stage mixing

3 Results and interpretations

Results of the mechanical tests in table (1) show that the two- stage mixing of concrete at early ages yielded significantly higher compressive strength than the conventional mixing concrete. However, there are no significant differences in the compressive and tensile strengths at 28-days between the two mixing techniques, although the two-stage mixing technique showed improved minimum and maximum micro-hardness. The micro-hardness curves at the age of 3 days and with w/c ratio of 0.45 for both mixing techniques are illustrated on the Figure 1. It can be seen that the two-stage concrete shows improvement in the minimum and maximum micro hardness over the conventional concrete. As the hydration proceeded from 3 to 28 days, the micro-hardness curves retained their overall shape. However the difference between

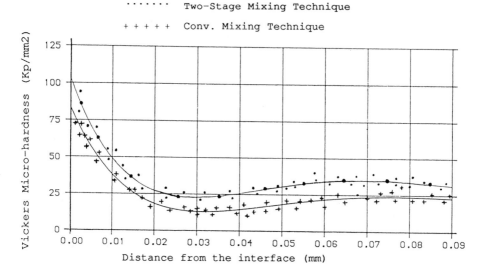

Fig.1 Micro-hardness at age of 3-day and 0.45 W/C ratio

the minimum and maximum micro-hardness tends to be less with age, as indicated on Figure 2.

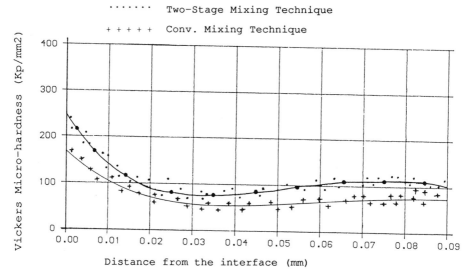

Fig.2 Micro-hardness at age of 28-day and 0.45 W/C ratio

By increasing the w/c ratio from 0.45 to 0.70, a similar

pattern was observed but the higher minimum and maximum values for the two-stage concrete were observed at all ages
including 28 days (Figures 3&4).

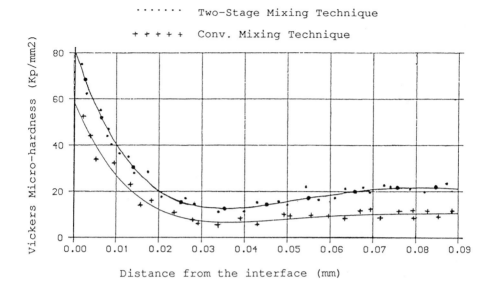

Fig.3   Micro-hardness at age of 3-day and 0.70 W/C ratio

The differences in the micro-hardness measurements between
the two mixing techniques correlate with the higher compressive and tensile strengths of the two-stage mixing concrete
particularly at early ages and high water/cement ratio.Since
the concrete made by both mixing techniques consists of the
same materials, any differences in the properties must be
attributed to differences in their microstructure.
Increasing the w/c ratio of the mix will result in increased porosity of the resulting concrete (Soroka 1968).
This reduces the strength of the concrete when mixed conventionally. The micro-hardness is also related to the porosity
(Beadoin 1975). The increased micro-hardness values observed
across the interface of the concrete made by the two-stage
process is therefore due to a lower porosity in this region
with more hydration products than in the conventionally
mixed concrete. This zone must therefore have a higher gel/
space ratio with a denser paste. Both the micro-hardness and
the compressive strength measurements involve the breaking
of similar bonds (Alexander 1968). It is reasonable to
suggest that in the concrete made by the two-stage process
the micro-hardness values reflect the breaking of more bonds
between the gel particles than for the more porous microstructure of the conventionally mixed concrete.

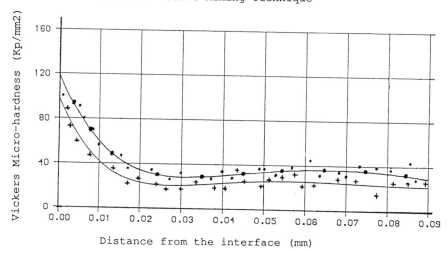

Fig.4 Micro-hardness at age of 28-day and 0.70 W/C ratio

4 Conclusions

Conventional mechanical tests have shown that up to 25%
higher compressive strength with significant decrease in
bleeding capacity at early age can be achieved by applying
the two-stage mixing technique.

The investigations also revealed that the two-stage mixing
concrete has resulted in higher minimum and maximum micro-
hardness measurements at the paste/aggregate interface. This
is attributed to the improvement of the bond at the cement
paste aggregate interface which is the weakest link in the
concrete, especially at early ages. It has been shown (Alexander 1968) that the paste strength is double that of the
bond strength at 7 days but that between 7 and 28 days the
bond strength gains steadily on the paste strength until the
failure mode changes from bond failure to the failure in
paste.

The increased micro-hardness of the concrete resulting
from the two-stage process can be attributed to the continuous filling of the porous interfacial zone with more hydration products, resulting in a higher gel/space ratio than
in the conventionally mixed concrete.

The resulting tighter packing of crystals with more intergrowth and interlocking means that the bond between the aggregate and cement paste has been increased and this results in the observed higher micro-hardness values.

5 References

Alexander, K.M., Wardlaw, J., and Gilbert, D.J. (1968) Aggregate-cement bond, cement paste strength and the strength of concrete. **The Structure of Concrete.** Ed. by A. E. Brooks and K. Newman, pp59-81, Cement and Concrete Association.

Beadoin, J.J. and Feldman, R.F. (1975) A study of mechanical properties of autoclaved calcium silicate system. **Cem. and Concr. Res.**, vol. 5, pp 103-118.

Carles-Gibergues, A., Grandet, J., and Ollivier, J.P. (1982) Contact Zone between Cement Paste and Aggregate. **Bond in Concrete.** pp 24-33, Ed by Bartos, P. Applied Science Publishers.

Hayakawa, M. and Itoh, Y. (1982) A new concrete mixing method for improving bond mechanism. **Bond in concrete.** PP282-288, Ed by Bartos, P. Applied Science Publishers, London.

Higuchie, Y. (1980) Coated-sand technique produces high strength concrete. **Concrete International.**

Maso, J.C. (1967) La Nature Mineralogique des Agregats, Fracteure Essentiel de la Resistance des Betons a la Rupture et al' action du Gel. **These,** Toulouse.

Mehta, P. K. (1986) **Concrete Structure, Properties and Materials.** pp 36-40, by Englewood Cliffs, 1986.

Soroka, I. and Sereda, P.J. (1968) Interrelation of hardness, modulus of elasticity and porosity in various gypsum systems. **J. Amer. Ceram. Soc.** 51(6)pp337-340.

5 SWEDISH METHOD TO MEASURE EFFECTIVENESS OF CONCRETE MIXERS

Ö. PETERSSON
Swedish Cement and Concrete Research Institute,
Stockholm, Sweden

Abstract
The Swedish regulations governing concrete structures: Materials and Workmanship, prescribe that the mixing time for concrete shall be at least 1,5 minutes. For cost reasons it has been asked whether this time might be reduced. A Swedish standard has been established for testing mixing time. This standard is based on a study at the Swedish Cement and Concrete Research Institute. The result of this study is presented in this paper. The findings were that the assessment may be made purely on the basis of the cement macro-homogeneity result which, according to the study, is the parameter which gives the best indication to the attainment of satisfactory mixing. The homogeneity is achieved when the coefficient of variation for cement is below 6%. The Swedish standard has been used for some years to reduce the mixing time for concrete mixers. The standard seems to perform satisfactory in practice. Case study will be presented.
Keywords: Mixing, Concrete Mixers, Cement variations, Macro-Homogenity, Micro-Homogenity, Effectiveness of Mixers.

1 Background

This first part of this paper is based on a study at the Swedish Cement and Concrete Research Institute, A Johansson (1971).

The 1965 edition of the Swedish Regulations Governing Concrete Structures: Materials and Workmanship - Concrete (BfB B5-1965) prescribes that the mixing time for concrete shall be at least 1.5 minutes (90 seconds). A wide variety of methods have been employed in efforts to determine the mixing capacity of concrete mixers. Most of these rely on measurements of the variations of one or more properties, for example:

    consistence
    unit-weight of concrete
    unit-weight of air-free mortar
    air content

Special Concretes: Workability and Mixing. Edited by Peter J. M. Bartos. © RILEM.
Published by E & FN Spon, 2–6 Boundary Row, London SE1 8HN, 0 419 18870 3.

content of fine and coarse aggregate
cement content
water content
compressive strength.

Since compressive strength is the property most sought for in concrete the determination of its variation is the most logical method. It suffers, however, from the disadvantage that strength takes a long time to determine and the testing procedure may cause wide scatter in the results.

The determination of variations in cement content and aggregate content has gained greatest favour as testing method. If the purpose of testing is to determine the shortest acceptable mixing time then the variation in cement content would appear to be the most suitable of these two criteria. It also appears to be the more sensitive than strength variation to shorter mixing time.

In the determination of distribution variations, the specimen must not be too small, since otherwise the presence or absence of a single stone in a sample would result in excessive variations. The specimens should therefore be at least 2 litres.

In scatter calculations of this type it is thus the variations between the 2-litre specimens that is obtained. This is termed macro-homogeneity. However, the variations in the distribution of the materials within each specimen - the micro-homogeneity - is still not known. In view of this, there must also be an upper limit to the size of the specimens. The number of specimens was fixed at 8 for this study.

From the practical viewpoint, the assessment of macro-homogeneity is of dominant significance. But the degree of micro-homogeneity may also have some influence on the properties of the concrete. The following criteria were adopted in the investigation:

1. Distribution of cement content within batch.
2. Distribution of fine aggregate within batch.
3. Distribution of coarse aggregate within batch.
4. Variations in compressive strength within the batch.
5. Change in compressive strength with increased mixing time.
6. Change in consistence with increased mixing time.

Criteria 1-4 are all expressions of macro-homogeneity, while 5 and 6 are - at least in part - expressions of micro-homogeneity. The relationship between homogeneity and mixing time can be affected by the following factors:

- Mix analysis
- Order of charging of ingredients
- Batch size, actual in relation to nominal
- Wear of mixer
- Cleanliness of mixer.

In view of the scope of the investigation it was not possible to vary all the above factors. Besides mixing time, the concrete mix analysis was included as a test variable in this study. The following three mixes were employed:

A. Cement content = 300 kg/m^3, consistence = T 5-10 cm, max. aggregate size = 36 mm

B. Cement content = 350 kg/m^3, consistence = P, 5-3 VB max. aggregate size = 16 mm

C. Cement content = 350 kg/m^3, consistence = SS, 20-10 VB max. aggregate size = 16 mm.

The investigation comprised three types of mixer:

I.   Horizontal type, 1500-litre batch.
II.   Gravity type, 2500-litre batch.
III. Horizontal type, 150-litre batch.

Each of these represents a common type of mixer. The latter is a laboratory mixer. In order to judge the properties of the mixer fairly, the specimens must be obtained during the normal course of the discharge process without affecting this. It was, for example, out of the question to stop the mixer for sampling.

The contents of cement, coarse aggregate and fine aggregate were determined by using the Pycnometer method. The method involves weighing in air and water, and the wet screening of a sample of concrete. At the same time, the contents of the other ingredients can be obtained through a somewhat extended wet screening process.

The decisive source of error in these calculations is the variation in the grading of the aggregate. One aim is to determine as accurately as possible the amount of aggregate less than 0.25 mm. Aggregate samples, as nearly as possible representative of the entire batch, were taken for aggregate used in a test batch of concrete.

As shown by Figure 1, referring to mixer I, the relationship between cement distribution and mixing time is distinctly dependent upon the type on concrete being mixed. Type A is a concrete having a high maximum aggregate size, high aggregate concrete, markedly gap-graded, and of relatively fluid consistence - all of which combine to make the fresh concrete easily segregated and unstable. The cement distribution results illustrate this. For short mixing times the cement is very unevenly distributed within the batch. Homogenisation improves relatively rapidly up to a mixing time of 60 seconds, after which it remains unchanged.

Type B represents a more stable concrete than type A due to its stiffer consistence, lower maximum aggregate size, lower aggregate content and higher cement content. This shows after a short mixing time, when the cement is somewhat better distributed. The improvement in homogeneity is not so rapid as for type A, probably because of the somewhat stiffer consistence. However, improvement continues instead to a better level.

Type C is clearly the concrete best suited to this type of mixer. The great stability of this concrete and the absence of any tendency to segregation result in a relatively high degree of homogenisation after only a short mixing time. Due to the stiffness of the mix, the improvement in homogeneity then proceeds slowly. The very low variations finally obtained are a sign that there are probably no deleterious effects due to layering and discharge segregation. As shown by the curve for type C concrete, variation due to testing methods is considerably less than 5%, the value at which the "A" curve levels off.

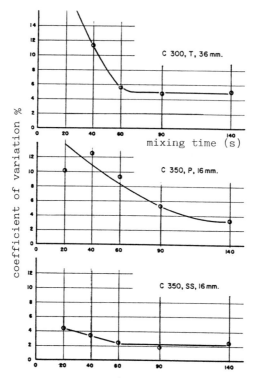

Fig 1.1500-litre horizontal mixer.
Variation of cement

In comparison with the 1500-litre horizontal mixer, the gravity mixer reveals certain differences as regards mixing ability, see Figure 2. A concrete of relatively fluid consistence (type A) is quickly mixed to a certain level of homogenisation, after which no further improvement occurs. This type of mixer is clearly better than the tested horizontal-type mixer when it comes to stiff mixes. On the other hand, the gravity mixer also reveals considerable deleterious effects due to layering and/or discharge segregation. In fact, there is even a mild tendency for the segregation of cement to increase with increased mixing time. As regards cement distribution, the type B concrete seems to be the best for the gravity mixer. The very small degree of segregation means that for a relatively stable mix the homogenisation will be good.

The small horizontal mixer (III) was tested only with concrete of types A and B, Figure 3. Type A proved unsuitable for this type of mixer. Considerably better results were obtained in the testing of type B. The speed of homogenisation is even more apparent here, where mixing time could be reduced to as little as 10 seconds without poorer cement distribution. It should be noted in this context that most of the ingredients have been stirred for 10-20 s before the actual mixing time starts see Figure 4.

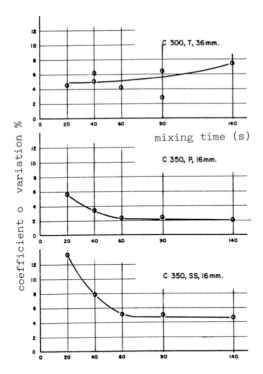

Fig 2. 2500-litre gravity mixer. Variation of
cement

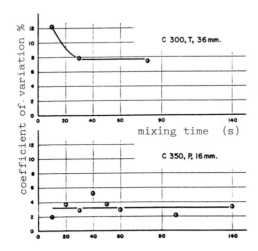

Fig 3. 150-litre horizontal mixer. Variation of cement

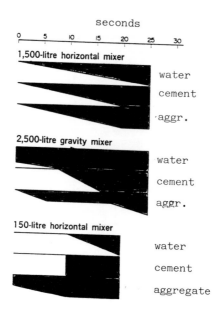

Fig 4. Premixing times

The constant level of the coefficient of variation for concrete of type B from the gravity mixer was as low as 2%, meaning that the variations caused by the test method must be less than 2%.

Even by visual inspection it was possible to determine that the micro-homogeneity of the concrete was not satisfactory at such short mixing times as 10-20 seconds. Some lumps of concrete were still dry inside and certain surfaces of the aggregate were not even wet. This makes it clear that even though full macro-homogeneity has been achieved, there is no certainty that the concrete is fully mixed.

The results for the aggregate show that it is distributed more quickly than cement, the coarse fraction being somewhat more quickly distributed than the fine.

The values for variations in strength, are given for various mixing times in Figure 5. A comparison with the corresponding curves for cement distribution variations shows roughly the same tendencies, whereas the coefficients of variation for strength are lower than for cement in the majority of cases. These lower coefficients may be due in part to the fact that strength variations are calculated on the basis of 8 values, each of which is the average for two test cubes.

Conclusions from the study were: The investigation reveals that the testing method employed is useful for the assessment of the homogenisation ability of a concrete mixer. The assessment may be made purely on the basis of the cement macro-homogeneity result, which is the parameter, which gives the best indication of satifactory mixing. An acceptable degree of homogeneity may be achieved when the coefficient of variation for cement content is below 6 %.

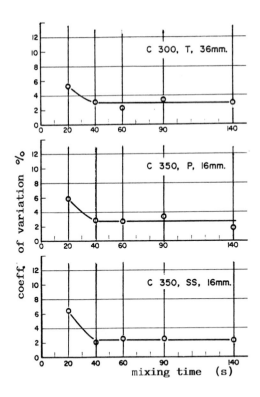

Fig 5. 2500-litre gravity mixer. Variation of compressive strength.

2 Swedish standard method

The result from the study were the base for the concrete mixers testing method. The method is described below:

Two different alternatives is allowed to use, a and b. For a, one has to use concrete of type 1, see Table 1, and for alternative b, one has to use concrete of type 2 and 3 see Table 1 .

Table 1. Concrete to be used in the tests.

Concrete type	Workability slump or VB	Cement content kg/m^3	Aggregate Ø max grading curve
1	Liquid, 100 - 150mm	300	38 mm, 1
2	Plastic, 20 - 50 mm	350	16 mm, 2
3	Very stiff, 10 - 20 s	350	16 mm, 2

* Grading curve see Fig 6 and 7.

For each concrete type three batches should be used and on each batch should at least three different mixing times be used from the following series: 30, 45, 60, 75 and 90 seconds. From each batch should 8 samples of 3 litre of concrete be taken. From each concrete type three batches should be mixed. The necessary mixing time is based on the maximum variation coefficient for cement according to Table 2.

Table 2. Limitations for cement variation.

Concrete type	Variation coefficient for cement
1	6 %
2	6 %
3	8 %

3 Tests on Mixer

Cement and Concrete Research Institute has performed investigations on different types of mixers in practise. The result shows that the standard is well suited and the limitations for cement variation coefficient is reasonable. Figure 8 shows an example from a test on a 7 m^3 gravity type of mixer (SERMEC type 6/7R MKII). The test was performed accordingly to alternative a, and measured at 45, 60 and 75 seconds. The result shows that the mixer has a lower variation coefficient than 6 % for the three mixing times used.

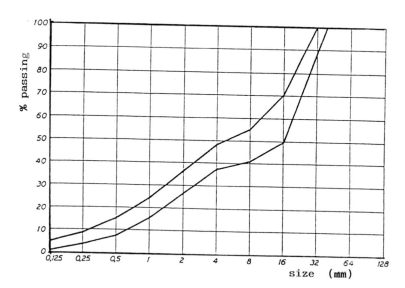

Fig 6. Grading curve for concrete type 1.

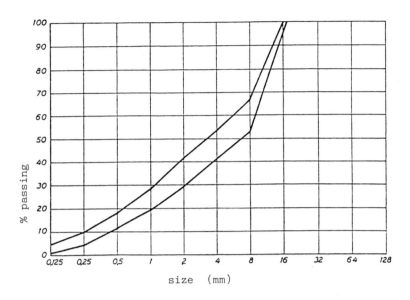

Fig 7. Grading curve for concrete type 2 and 3.

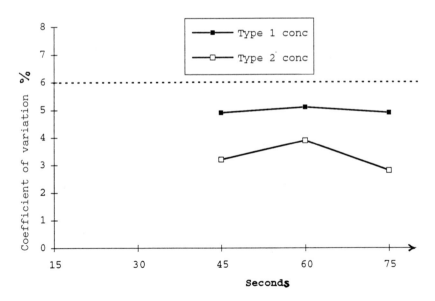

Fig 8. Test on 7 m^3 gravity mixer 1991. Variation of cement.

One can say that the method earlier describe is well suited for the macro-homogenity but for the micro-homogeneity we have to use other methods. When using high performance concrete with additives, like silicafume, then one has to look the micro-homogeneity. One method that can be used is thin-section techniques. In one project in progress at CBI, we study high performance concrete and the ability to mix it and in which order the ingredients should be mixed. The analyse method is to use thin-section techniques to look into the homogeneity in small scale.

4 References

A Johansson, The Relationship Between Mixing Time and Type of Concrete Mixer, The Swedish Cement and Concrete Research Institute, Proceedings NR 42 Stockholm 1971.

PART THREE

WORKABILITY AND MIX DESIGN

6 WORKABILITY AND RHEOLOGY

C. LEGRAND
Laboratoire Matériaux et Durabilité des Constructions,
INSA-UPS, Toulouse, France

Abstract
The heterogeneity of fresh concrete is an important
hindrance to the global rheological characterization of
this material and the mechanical interaction between phases
must be taken into account if a theoretical modelisation of
the behaviour is needed. In the practice, the difficulty
can be evaded using workability tests that would simulate
as much as possible the placing conditions.
Keywords : Coaxial viscometer, Heterogeneity, Interactions,
Placing, Rheological behaviour, Workability tests.

Before Poiseuille laid the foundations for modern
viscometry, ignorance of constitutive equations and related
characteristics led to fluids aptitude to flow being
described in subjective and pictorial terms like "thick",
"heavy", "tenuous", "sticky", "gluey", etc. As things
stand, it has to be admitted that we have not made much
progress since these former considerations when we describe
concrete as being "thick", "stiff", "firm", "plastic",
"fluid", "workable", etc.

The formalisation of fluids rheological behaviour, the
expression of the corresponding laws and the measurement of
the specific characteristics has allowed us to replace this
entire vocabulary with physical magnitudes, like the
coefficient of newtonian viscosity for example, which,
scientifically speaking, represents considerable progress.
We merely need to know, for example, that an oil has a
viscosity coefficient of 2.3 Pa.s at 20° C for it to be
completely identified rheologically at that temperature,
without needing to specify with what instrument or what
operating mode that value was measured. But we have to
admit that, as far as fresh concrete is concerned, we have
not yet reached this stage and that major difficulties
arise as soon as we try to attribute objective rheological
behaviour to this material.

Special Concretes: Workability and Mixing. Edited by Peter J. M. Bartos. © RILEM.
Published by E & FN Spon, 2–6 Boundary Row, London SE1 8HN, 0 419 18870 3.

A number of research scientists have tried to apply the same experimental process as that used with newtonian fluids to fresh concrete. Thus, as an example, coaxial viscometers of extremely large dimensions, considering the sizes of the larger aggregates, have been built, and curves have been plotted giving torque as a function of angular velocity. Unfortunately, the results obtained prove to be disappointing when we try to translate them into rheological behaviour laws obeying Noll general principles. The main reason for this is simple : fresh concrete is a highly heterogeneous material, with extreme internal mechanical discontinuities, and assuming medium continuity is unacceptable. This leads to the very notions of stress and gradient, which are inseparable from the continuous medium concept, being inapplicable in these instances.

Thus, while torque is indeed a mechanical value (depending on the material being studied, but also the equipment's geometry), any computation intended to transform torque into stress becomes meaningless. The same applies if we try to obtain a shear strain rate from the angular velocity.

Certainly, we can always force the medium to flow in a coaxial viscometer, but the results for measurements carried out in these conditions can by no means be translated by a flow curve and thus cannot allow intrinsic rheological magnitudes to be obtained. We did, in fact, transform a viscometer into a workabilimeter, but this was devoid of real interest as the type of flow obtained was remote that which exists in reality.

It is therefore indispensable to take the heterogeneity of fresh concrete into account if we wish to study its rheological behaviour. Unfortunately, the phases present are numerous : solid, free liquid, adsorbed liquid and gaseous ; further, the dimensions, the nature and the shape of the grains vary tremendously. Obviously, it would be unthinkable to consider fresh concrete in such a way as to take into account its overall heterogeneity. Rather a simpler model must be elaborated.

We have already tried (Barrioulet and al., 1975) to modelise vibrated fresh concrete by a diphasis medium comprising :
- the interstitial paste with its fluid flow behaviour,
- the granular matter whose shearing strength is the result of complex interactions involving contacts, shocks and solid frictions.

Using this model, we tried to understand whether the characteristics chosen as being suitable for each of these phases was adequate to determine fresh concrete's behaviour. In other words, where :

R_1 represents the range of rheological characteristics describing the paste's behaviour (Legrand, 1972),
R_2 represents the range of physical characteristics describing the aggregate's behaviour (Barrioulet and al., 1978),
C represents the vibrated concrete's behaviour in a given flow (in the LCL workabilimeter, for example).

In this conditions, two pastes of a different nature but having the same R_1 characteristics mixed in the same proportion with the same aggregate or with another of different nature but having the same R_2 characteristics should result in two concretes with the same C behaviour.

All the tests we managed to carry out showed that this was not the case and we were able to highlight the fact that in addition to the considerable difference in characteristics between the paste and the aggregates, the interaction between the two phases could not be reduced to a simple question of lubrication but to a particularly complex combination between the viscous effects of the paste and the mass of aggregates (Barrioulet and al., 1990).

The difficulty represented by rheological study of this material can be fully appreciated, but clearly an attempt to mask the level of complexity introduced by consideration of heterogeneity through carrying out global measurements on the concrete is devoid of any objective significance.

But this does not mean to say that such an approach is useless. Quite the contrary, in practical terms, this approach leads to workability tests, which constitute the only way of obtaining essential data on the capacity of the material to offer the performance expected of it, directly and simply. It is important no to seek to obtain global and intrinsic rheological characteristics through these tests and to bear in mind that the results will, of course, depend on the material studied but also on the equipment geometry and the test conditions.

Designing a good workability test must above all be guided by the concern to put the concrete to be characterised in conditions resembling as much as possible the placing conditions, and measuring the parameters which are most closely linked to the sought for qualities. Finally, it should be borne in mind that correlations with other tests are often random, except where the fresh concrete is gauged by the eye of an experienced worker who always has a clear idea of the workability of the concrete he will have to place as it is turned in the mixer.

References

Barrioulet, M. and Legrand, C. (1975) Influences
respectives de la pâte et des granulats sur la mesure
des caractéristiques rhéologiques du béton frais vibré.
Cahiers du Groupe Français de Rhéologie, Tome III, n°
5.
Barrioulet, M. and Legrand, C. (1978) Etudes des
frottements intergranulaires dans le béton frais. Idées
nouvelles sur l'écoulement du béton frais vibré.
Matériaux et Constructions, Vol. 11, n° 63, 191-197.
Barrioulet, M. and Legrand, C. (1990) Les interactions
mécaniques entre pâte et granulats dans l'écoulement du
béton frais. **Proceedings of the Congress "Properties of
Fresh Concrete" (Hanover)**, Chapman and Hall, 263-270.
Legrand, C. (1972) Contribution à l'étude de la rhéologie
du béton frais. **Matériaux et Constructions**, Vol. 5, n°
29, 275-295.

7 WORKABILITY AND WATER DEMAND

D. W. HOBBS
British Cement Association, Crowthorne, Berkshire, UK

Abstract
In this paper the main factors influencing the workability
and water demand of Portland cement concretes and
concretes in which part of the Portland cement or part of
the aggregate has been replaced by fly ash or slag is
discussed. It is proposed that the main parameters
influencing workability are the volume concentration of
cementitious particles in the paste fraction and the
volume concentration of aggregates particles in the
concrete.
Keywords : Flow, Workability, Water Demand, Bingham Model,
Portland Cement, Fly Ash, Slag, Slump, Compacting Factor.

1 Introduction

The objective of concrete mix design is to select the
constituent materials, crushed fines or natural sand,
coarse aggregate, Portland cement or composite cement and
admixture, and to proportion the various constituents to
produce concrete of the correct workability that, when
hardened, is fit for its intended purpose. In addition,
it must be possible to produce the concrete at a
reasonable cost.

The specification for the concrete may require the
following limiting values to be met: a characteristic
compressive strength, a maximum water/cement ratio, a
minimum cement content and a specified content of entrained air. The specification may also give the type of
cement, the required workability, the maximum aggregate
size and the rate of sampling for strength testing. The
specification forms the objective of the mix design
process.

Many factors are known to influence the properties of
concrete including: aggregate type, maximum size, grading
and variability, Portland cement type and variability, fly
ask or slag type and variability, air content and variability, the mix proportions, the quality of the mixing
water, mixing, compaction, segregation, curing and time.

Special Concretes: Workability and Mixing. Edited by Peter J. M. Bartos. © RILEM.
Published by E & FN Spon, 2–6 Boundary Row, London SE1 8HN, 0 419 18870 3.

The manner and extent to which these various factors affect strength and workability is not clearly known or understood, consequently optimisation of the mix proportioning of a concrete is not possible without trial mixes. When designing the initial trial mix only factors having a major influence upon compressive strength and workability are taken into account. This is the basis of most methods of mix design. In this paper the influence of a number of factors upon workability and water demand are considered.

2 Flow behaviour

The flow behaviour of a cement paste is complex and published data show no general agreement on this behaviour [Helmuth(1980)]. The flow properties of cement paste have most commonly been determined by using the coaxial-cylinders viscometer method [Hobbs(1980)]. Figure 1 shows the parameters involved. The viscosity of a homogeneous fluid can be readily measured by using this test; however, when the fluid is a particle suspension, such as a cement paste, the values obtained will not be fundamental material constants but will be influenced by the test conditions. If the character of the steady-state flow is not influenced by solid particle concentration and shear rate, the test will give an indication of the changes in 'viscosity' produced by changes in water/cement ratio and shear rate.

Cement pastes in concretes typically have water/cement ratios ranging from 0.4 to 0.8 and the flow properties of such pastes have been determined by Ish-Shalom and Greenberg(1962), Odler, Becker and Weiss(1978) and vom Berg(1979). The cement pastes tested by these workers showed flow behaviour which could be represented approximately by the Bingham model, thus

$$\tau_p = \tau_{op} + \mu_p \, \dot{\gamma}_p \qquad\qquad (1)$$

where τ_p is the applied shear or viscous stress, μ_p is the plastic viscosity, $\dot{\gamma}_p$ is the velocity gradient or shear rate and τ_{op} is the yield value of the cement paste. τ_{op} is caused by interparticle forces and interlocking of the 'flocculates'.

Rheological measurements on concrete are more difficult than on pastes because of the larger particle sizes of the aggregates, but attempts have been made to determine flow behaviour by using a coaxial viscometer with a gap width only 1.6 times [Uzomaka(1972)] and 2.5 times [Murata and Kikukawa(1973)] the maximum aggregate particle size. On the basis of these tests, it has been suggested that the flow behaviour of concrete, like that of its cement paste, approximates to the Bingham model [Uzomaka(1972), Murata and Kikukawa(1973), Tattersall(1991) and Morinaga(1973)].

The flow behaviour of concrete will be at least as complicated as that of its cement paste but, in the author's view, the flow properties of the cement paste fraction in a concrete may be essentially the same as those determined on similar cement paste samples, provided the shear rate is similar in each case. For this to be so,

$$\dot\gamma_p = \frac{\tau_p - \tau_{op}}{\mu_p} \approx \dot\gamma_c(1 - V_a) \tag{2}$$

where $\dot\gamma_c$ is the bulk velocity gradient of the concrete and V_a the aggregate volume concentration.

At high rates of shear, and provided streamline flow occurs, the Bingham equation for concrete reduces to

$$\tau_c = \mu_c \, \dot\gamma_c \tag{3}$$

when $\tau_{qc} \le \mu_c \, \dot\gamma_c$. This is the rheological equation of a Newtonian fluid where μ_c is termed the fluid viscosity.

It follows that a single-point test, and this is what is generally used, may provide useful information regarding the 'workability' of concrete if
(a) a high rate of shear occurs both in the single-point test and when the concrete is placed on the site; or
(b) the same rate of shear occurs both in the single-point test and when the concrete is placed on the site [Nessim and Wadja(1965)] or
(c) the yield value for concrete, τ_{oc}, is broadly related to its plastic viscosity, μ_c.

3 Workability and water demand. Portland cement concretes

When coarse or fine particles are added to a concrete, mortar or cement mix, then irrespective of their shape or surface texture, the workability of the mix will be reduced because of the increased viscous drag provided by the particles. Concrete will become unworkable when either the cement particles or aggregate particles are in close contact [Hobbs(1976 and 1981)]. In order for the paste to be workable it is necessary to add sufficient water both to fill the voids and to ensure that the cement particles are not in close contact. The further the cement particles are apart, or alternatively the lower the volume concentration of the cement, the more workable the paste. Similarly, in order for concrete to be workable, it is necessary to add sufficient paste both to fill the voids between the aggregate particles and to ensure that the aggregate particles are not in close contact. The further the aggregate particles are apart, or alternatively, the lower the volume concentration of the aggregate, the more workable the concrete.

It is often assumed that the surface area of the particles largely controls workability and water demand. In Figure 2 and Table 1 it is shown that this assumption is not justified.

Table 1 gives the slump and compacting factor for a number of concrete mixes in which one or other of the four sand gradings given in Table 2 was used. In these tests only sand retained on or above a BS 150 μm sieve was used. No obvious trend in workability with both increasing fineness and increasing proportion of sand is apparent from the results.

Table 1. Slump and compacting factor

| w/c | a/c | slump (mm) | | | |
| | | sand grading* (% sand) | | | |
		1 (30)	2 (35)	3 (42)	4 (48)
0.35	2.90	65	60	70	50
0.35	3.30	20	25	40	20
0.35	4.37	0	0	0	0
0.47	4.46	h	h	75	65
0.47	5.15	5	20	25	15
0.47	8.17	u	15	10	5
		compacting factor			
0.35	2.90	0.87	0.91	0.95	0.88
0.35	3.30	0.81	0.83	0.86	0.78
0.35	4.37	0.75	0.76	0.75	0.75
0.47	4.46	0.90	0.94	0.95	0.97
0.47	6.01	0.84	0.86	0.86	0.84
0.59	8.17	u	0.82	0.90	0.87

* see Table 2, u unworkable, h collapse

Figure 2 shows a comparison of the effect upon slump of replacing various proportions of combined aggregate by equal volumes of slag or fly ash. Increasing the surface area of the particles does not necessarily reduce slump; also, although the slag particles are irregular and the fly ash particles spherical, the effect of these fine materials upon changes in slump is similar. Replacing up to 5 percent by weight of the aggregate by slag or fly ash increases workability because the increase in workability produced by the reduction in aggregate volume concentration is greater than the reduction produced by the addition of slag or fly ash. At higher aggregate replacement levels, the slump declines because the

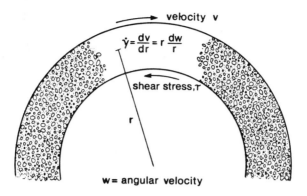

velocity v

$$\dot{\gamma} = \frac{dv}{dr} = r\frac{dw}{r}$$

shear stress, τ

r

w = angular velocity

Fig. 1. Coaxial-cylinders viscometer method.

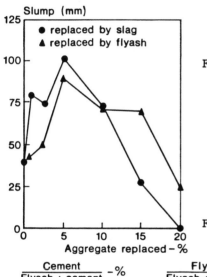

Slump (mm)

● replaced by slag
▲ replaced by flyash

Aggregate replaced – %

Fig. 2. Effect upon slump of replacing various proportions of the aggregate by slag or fly ash. w/c 0.77

Fig. 3. Boundary between workable and un-workable mixes.

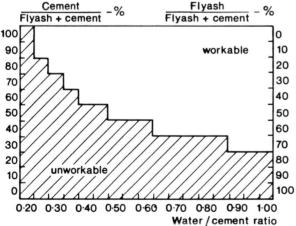

$\frac{\text{Cement}}{\text{Flyash} + \text{cement}}$ – % $\frac{\text{Flyash}}{\text{Flyash} + \text{cement}}$ – %

workable

unworkable

Water / cement ratio

increase in workability produced by the reduction in aggregate volume concentration is now less than the reduction in workability produced by the addition of fly ash or slag. The concrete becomes unworkable when the cement and fly ash particles or cement and slag particles are in close contact.

Table 2. Sand gradings

Sieve size	Percentage by mass passing BS sieve			
	1	2	3	4
5.0 mm	100	100	100	100
2.36 mm	77	80	93	87
1.18 mm	53	60	67	71
600 μm	30	40	50	56
300 μm	7	8.5	12	25
150 μm	0	0	0	0

If fine sand of similar size to the cement particles is included in the mix the concrete is unworkable when either particles of cement size or coarser aggregate particles are in close contact. Thus the workability is some function of

$$V_{am} - V_a \text{ and } V_{cm} - V_c$$

where V_{am} and V_{cm} are the volume concentration of aggregate particles and cement sized particles which produce an unworkable mix and V_a and V_c are the actual volume concentration of aggregate and cement sized particles.

Reducing the maximum aggregate particle size or changing from an uncrushed to crushed rock aggregate will reduce V_{am} and will, therefore, reduce workability. Consequently, water content will need to be increased to maintain workability. Increasing the proportion of fine sand passing a BS 150 μm sieve will reduce $V_{cm} - V_c$ and, consequently, water content will need to be increased to maintain workability.

The main factors influencing water demand are therefore: specified workability, maximum aggregate size, aggregate type, crushed or uncrushed, aggregate grading, aggregate volume concentration and cement volume concentration.

When the cement content of a mix is increased, the workability is normally maintained by reducing both the aggregate content and the proportion of fine to coarse aggregate.

4 Workability and water demand. Portland/slag and Portland fly ash concretes

When a fine material of lower reactivity than the Portland cement, such as fly ash or slag, is added as a partial replacement by volume for the Portland cement, then the workability will be increased because the increase in workability resulting from the increase in water Portland cement ratio is greater than the reduction caused by the added material [Hobbs(1980 and 1981)]. Thus the use of fly ash and slag as partial replacements for Portland cement would be expected to reduce water demand, the extent of the reduction depending upon the volume concentration of Portland cement in the paste fraction. This is illustrated in Figure 3 which shows the boundary between workable and unworkable mixes [Hobbs(1981)]. The volume concentration of solids (Portland cement plus fly ash) which produced an unworkable mix was approximately 0.58 which was essentially independent of water-Portland cement ratio.

In Figure 4, the changes in water demand are shown for concretes in which the binder contains 30 or 35 percent fly ash relative to Portland cement concretes of similar slump and 28 day compressive strengths. The water reduction decreases with increasing Portland cement content becoming negligible at about 450 kg/m^3 [Hobbs(1988)]. The actual changes in water demand depend upon the chosen cements and ashes. For mixes with binder contents from 275 to 400 kg/m^3 it has been found that the use of a fly ash which complies with BS 3892: Part 1 (1982) generally results in a reduction in water demand of between 7 and 12 percent. This is in broad agreement with the U.K. Department of Environment method of mix design [Teychenné, Nicholls, Franklin and Hobbs(1988)] which assumes a reduction of 3 percent for each 10 percent fly ash in the binder.

If the Portland cement content remains unchanged and the aggregate is partially replaced by fly ash or slag, then the effect upon workability is less easy to predict - the reduction in aggregate volume will increase workability but its partial replacement by fly ash or slag will reduce workability, the extent of the reduction depending upon the volume concentration of Portland cement in the paste fraction.

A comparison of the effect upon slump, compacting factor and flow of replacing aggregate by an equal volume of slag, fly ash or Portland cement is shown in Figures 5 to 7. The effect of these fine materials is broadly similar. Portland cement produces mixes of the lowest workability possibly because it both reacts with a proportion of the water increasing the volume concentration of solids and reducing free water content. Slag produces mixes with workabilities intermediate between

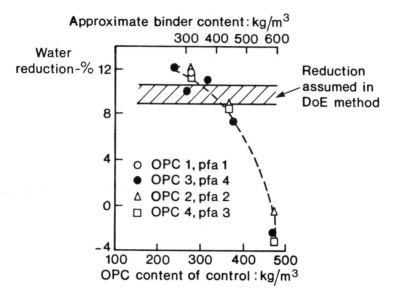

Fig. 4. Reduction in water demand and PC content of control; constant slump and 28 d compressive strength (30 or 35% fly ash).

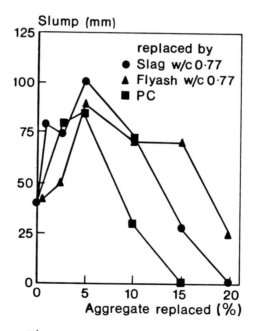

Fig. 5. Effect upon slump.

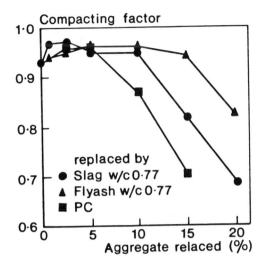

Fig. 6. Effect upon compacting
factor.

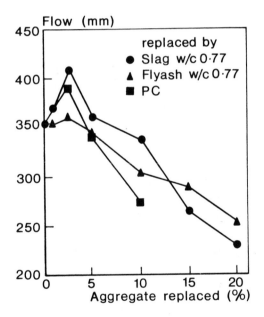

Fig. 7. Effect upon flow.

those in which the aggregate is partially replaced by Portland cement or fly ash.

5   **Acknowledgements**

The author and publishers are grateful to Thomas Telford Publications, London for allowing them to reproduce Figure 4.

6   **References**

British Standards Institution (1982) Pulverized-fuel ash. Part 1 Specification for pulverized-fuel ash for use as a cementitious component in structural concrete. BSI London, BS 3892: Part 1.

Helmuth, R.A. (1980) Structure and rheology of fresh paste.Proc 7th Int. Congress on the Chemistry of Cement, Paris, 1980. Paris, Editions **Septima**, 1980. III, VI/O, 16-30.

Hobbs, D.W. (1976) Influence of aggregate volume concentration upon the workability of concrete and some predictions resulting from the viscosity-elasticity analogy. **Mag. Conc. Res.**, 28, 191-202.

Hobbs, D.W. (1980) The effect of pulverized-fuel ash upon the workability of cement paste and concrete. **Mag. Conc. Res.**, 32, 219-244.

Hobbs, D.W. (1981) Discussion of Hobbs (1980) **Mag. Conc. Res.**, 33, 227-244.

Hobbs, D.W. (1988) Portland-pulverized fuel ash concretes: water demand, 28 day strength, mix design and strength development. **Proc. Inst. Civ. Engrs.**, Part 2, 85, 317-331.

Ish-Shalom, M. and Greenberg, S.A. (1962) The rheology of fresh Portland cement pastes. Proc. 4th Int. Sym. on the Chemistry of Cement, 2-7 October 1960, Washington, D.C. Washington, D.C. National Bureau of Standards. Monograph 43, II, 731-748

Morinaga, S. (1973) Pumpability of concrete and pumping pressure in pipelines. Fresh concrete: Important properties and their measurement. Proc. RILEM Seminar, 22-24 March, 1973, Leeds. Leeds, The University, 3, 7.3-1 to 7.3-39.

Murata, J. and Kikukawa, H. (1973) Studies on rheological analysis of fresh concrete. Fresh concrete: Important properties and their measurement. Proc. RILEM Seminar,22-24 March, 1973, Leeds. Leeds, The University, 1, 1.2-1 to 1.2-33.

Nessim, A.A. and Wajda, R.L. (1965) The rheology of cement pastes and fresh mortars. Mag. Conc. Res., 17, 57-68.

Odler, I. Becker, T. and Weiss, B (1978) Rheological properties of cement pastes. Il Cemento, 75, 303-310.

Tattersall, G.H. (1991) Workability and quality control of concrete, London, E&F N Spon.

Teychenné, D.C., Nicholls, J.C., Franklin, R.E. and Hobbs, D.W. (1988) Design of normal concrete mixes. Department of the Environment, BR 106.

Uzomaka, O.J. (1972) A concrete rheometer and its application to a study of concrete mixes. Proc. 6th Int. Congress of Rheology, Lyons, 4-8 September 1972. Darmstadt, Dr Dietrich Steinkopft Verlag, 4, 233-235.

Vom Berg, W. (1979) Influence of specific surface and concentration of solids upon the flow behaviour of cement pastes. Mag. Conc. Res., 31, 211-216.

PART FOUR

FIBRE CONCRETE

8 PROPORTIONING OF FIBRES AND MIXING OF FIBRE CONCRETE

N. H. NIELSEN
Skako A/S, Faaborg, Denmark

Abstract
The paper considers the mixing plant for production of special concretes. A particular focus is on the proportioning and mixing of steel and polypropylene fibres. Factors which influence efficiency of the mixing plant designed to deal with fibre-concretes are discussed.
Keywords: Concrete Mixers, Batching Plant, Steel Fibres, Polymer Fibres, Mix Design.

1 Introduction

Skako A/S has been involved in the manufacture of plant for production of different special concretes for a number of years. The special concretes have included:

Very dry, precasting, concrete.
(Water control monitored by an ohmmeter)

Concrete containing different types of fibres.

Micro-silica concrete.
(eg. for the Great Belt bridge project)

High strength concrete.
(eg. the Densit type)

Flyash concrete.

Lightweight concrete.
(Inorganic and polystyrene aggregate)

The experience obtained suggests that a successful solution of the problem of mixing some of the special concretes has to begin with the storage and proportioning systems.

This is not only because the storage and proportioning (batching) of the special constituents of such mixes may be itself difficult but also because the manner in which these constituents are added into the mixer may have a considerable influence on the mixing process as a whole.

Special Concretes: Workability and Mixing. Edited by Peter J. M. Bartos. © RILEM.
Published by E & FN Spon, 2–6 Boundary Row, London SE1 8HN, 0 419 18870 3.

This can influence significantly the necessary mixing time and the overall mixing efficiency of the plant.

2 Batching and mixing

The fibres are normally supplied either pre-weighed in portions specified by the customer or as bulk goods weighed manually prior to batching.

When fibres are considered, several factors can make the mixing of the fibres into the concrete difficult, namely:

(a) The tendency of the fibres towards entanglement. This is mainly a function of the type/dimensions of the fibres.

(b) The difference between density of the fibres and the density of fresh concrete. This makes it often necessary to choose the correct time in the mixing cycle for the addition of the fibres. A typical example is the addition of lightweight polymeric fibres. The addition before water is much better because the addition after water is into fresh liquid concrete and the buoyancy of the fibres makes the mixing difficult.

(c) The changed character of the fresh mix following the addition of fibres. There is little change in the mixing process when small quantities of fibres are added (eg. < 1kg of polymer fibres per $1m^3$ of concrete). However, trial mixes using the actually selected mixer are necessary when large quantities of fibres are added.

Manual batching continues to be used in cases of fibres with which it appears impossible to avoid entanglement when using mechanical systems. The entanglement can cause problems both at the storage and at the batching stages of the production process.

The manual batching has a number of drawbacks:

- An extended mixing time. This may be necessary in order to achieve a homogeneous mix. It can be also due to an uneven manual batching. A homogeneous mix is one which contains an identical quantity of fibres in each part and that these fibres are separated, i.e. they do not appear in bunches.
- A possible need for an unnecessarily more expensive mixer in order to achieve the desired homogeneity and the possible under-utilization of the capacity of such plant.
- A substantial time consumption for the weighing and proportioning.
- An insufficient control of the whole batching process.
- A lack of documentation for the control of the quantity of fibres actually added.

Mechanical batching plant which avoids the drawbacks mentioned above has been developed for steel fibres and polymer fibres.

2.1 Steel fibres
The steel fibre batching equipment produced by Skako consists of two vibratory feeders connected in series (Fig.1). The first vibratory feeder works both as a silo for the material and as the batching feeder. The second vibratory feeder works as a weighing hopper and as

a proportioning feeder respectively, with a levelling function for the specifically weighed charge to be added to the concrete batch. The second vibratory feeder evenly proportions the weighed quantities of fibres to the position in the process where the steel fibres are required.

The steel fibres can be batched directly into the concrete mixer, the belt conveyor, the weighing hopper or into the truck mixer. The proportioning is controlled by a centrally located process control system which ensures continuous supervision and an optimum utilization of the plant.

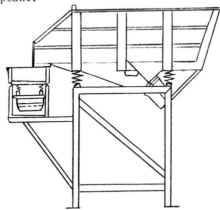

Fig.1 Steel fibre batching equipment.

The advantages of using mechanical batching of steel fibres are:
- An automated proportioning with a connection to the process control system, thus ensuring sufficient control and providing necessary documentation.
- A guarantee that the batched quantities are correct. Accurate weighing.
- A continuous, fast batching of fibres directly to the point of mixing which helps to achieve the best distribution of the fibres in the mix.
- A reduction of manual labour requirement.
- Improved working conditions in an automated plant with no heavy manual labour involved.
- Suitable for bulk fibre supplies which reduce the costs.
- A compact plant which can be tailored to individual customer's requirements.

2.2 Polymer fibres
The polymer fibre batching plant consists of the following main parts:
Fibre cutting and batching unit with a large storage space for coils of the polymer fibre cord (Fig.2).
Reeling and surveillance equipment.
Cutting unit with a selector of fibre lengths from 6 to 30 mm.
Control unit either as a separate process control or as an integral

part of the overall process controller.
The batching is carried out on a time basis (volumetric). The capacity is adapted to the specific requirements of the production.
The batching plant is a compact unit generally suitable for installation directly in a specific place in the production plant where the polymer fibre batching is required.
An automatic gate which is open only during the batching is located at the outlet from the fibre cutter.

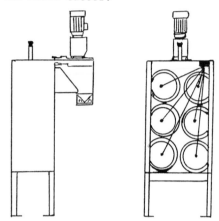

Fig.2 Polymer fibre cutting and batching plant.

The advantages achieved by using special mechanical plant for batching and dosage of polymer fibres are similar to those for batching of steel fibres mentioned before (2.1) with a further advantage of being able to use fibres supplied in coils thus reducing the costs.

3 Conclusions

Highly automated, efficient mechanical batching systems are available for a variety of special concretes, particularly for mixes with fibres. However, it is still necessary to consider each type of fibre and type of mix individually and adjust the batching and mixing process to produce the best quality concrete mix with the least costs. It is important to match the batching / proportioning equipment with the mixer and decide on the correct timing of the addition of all the concrete constituents into the mixer.

9 MIXING OF GLASS FIBRE REINFORCED CEMENT

I. D. PETER
Power-Sprays Ltd, Bristol, UK

Abstract
Glass Reinforced Cement differs from standard concrete in that it is reinforced with Alkali Resistant Glass Fibre, has a high cement content and contains no aggregate above 1mm in diameter. Because of the nature of the mix special types of mixers have been developed and special test methods devised for the assessment of workability and homogeneity of GRC mixes.
Keywords: Glass Fibres, GRC, Cement Slurry, Premix, Roving, Workability, Alkali Resistant Glass Fibre, Mixers, Washout, Mini-slump.

1 Alkali resistant glass fibre

Alkali Resistant Glass Fibre can be supplied as a Roving or as Chopped Strands. The fibre imparts toughness to the finished composite but it is itself brittle and can be easily damaged during the mixing process. A glass fibre roving consists of 20 to 50 strands of fibres each containing between 50 to 200 individual filaments. The strands are wound together to form the roving. The integrity of the strand must be maintained during the production of the GRC because the breakdown of the strands into individual filaments will drastically affect the workability of the fresh mix and reduce the mechanical mproperties of the cured composite.

2 Manufacturing process

There are two basic methods in common use for the manufacture of GRC components. In the first of these processes, **the spray process,** a cement slurry is mixed and then pumped into a spray gun which chops the glass fibre strands and simultaneously sprays the slurry and the chopped fibres onto or into a mould. The spray gun , Fig. 1, can be hand-held or a part of a fully or semi-automated process.
    The second process is used in vibration casting where the chopped glass fibre strands are added into the cement slurry during the mixing process. This method is normally referred to as the **premix process.**

Special Concretes: Workability and Mixing. Edited by Peter J. M. Bartos. © RILEM.
Published by E & FN Spon, 2–6 Boundary Row, London SE1 8HN, 0 419 18870 3.

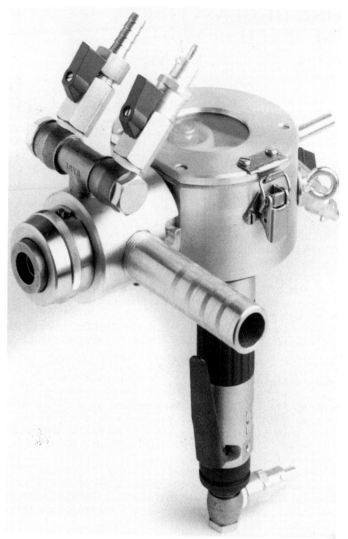

Fig. 1. A concentric glass fibre reinforced cement (GRC) spray gun.

3 Mixing and mix design

The cementitious slurry is produced using a 'High Shear Mixer',
Fig. 2., with the mixing tools revolving at approx. 1400 rpm. The **high
shear method** of mixing is chosen to produce workable mixes having
high cement contents and low water cement ratios.

Fig.2. A high shear cement slurry mixer.

A typical batch of a cement slurry has the following composition:

    cement OPC / RHPC   ....   50 kg
    silica sand 0.25-0.75 mm .....   50 kg
    water ....   17 *l*
    superplasticizer ....   500 m*l*

The proportions shown above represent a mix with a w/c of 0.34 and a cement content of 900 kg/m3. The lack of any aggregate greater than 1 mm maximum particle size means that there are no large stones present which can assist in dispersing the cement particles and breaking up any lumps.

This is compensated for by the mixing at high speed with a specially designed blade. This creates a vortex effect which ensures a rapid dispersion od cement and an efficient mixing. The sequence in which the mix constituents are added is important. Water with the admixture are added first, followed by the cement and finally the sand. The slurry produced in this manner is used in the **spray-process** previously described.

Adding glass fibres into this mixing process would result in the breakdown of the glass fibres strands and it is therefore necessary to use a different mixing technique when the **premix** process is considered.

It is possible to transfer the slurry produced in a high shear mixer into a small pan-type mixer (eg. Cretangle, Cumflo etc.), add manually the pre-weighed quantity of the pre-cut glass fibre strands and mix them in. This method can produce an acceptable **premix GRC** but it has the disadvantage of using two mixers, doubling the cleaning and maintenance requirements, requiring a supply of pre-cut fibre and being difficult to automate and avoid manual batching of the fibres.

New mixers have been developed which can produce both the high speed, high shear mixing necessary to produce a good quality cement slurry and the gentle blending action required for the incorporation of the glass fibres.

A mixer shown on Fig. 3 works at high speed when mixing together the cement, water, sand and admixtures and changes to a low speed mixingto blend in the glass fibres. Automated versions such as the one shown on Fig.3 also feature weigh hoppers fed by screw conveyors for an accurate batching of cement and sand. Water and admixtures are measured volumetrically and the glass fibre is dispensed via a multi-roving fibre chopper (Fig.4) fitted with a timer which regulates the total weight of the fibre added.

The workability of the fresh premix is of prime concern with the consistency of the slurry being of the essence. The addition of the glass fibres reduces the floe properties of the slurry and this must be taken into account when calculating the original mix design. Once the glass fibres have been added it is not possible to add more water or plasticizer to change the workability. Attempts at doing so will lead to the fibre separating out from the mix and a tendency to an increased bleeding during later processing.

4 Control of the mixing process

Uniformity of the mix can be obtained by :
   a. Accurate weighing of dried ingredients
   b. Automatic dispensing of water and additives
   c. Preset or automatically controlled mixing times

Fig.3. A premix-GRC mixer.

5 Tests for assessment of workability and homogeneity of the GRC mix

5.1 Mini-slump test (GRCA)
This test uses a plastic plate 300 mm square on which concentric
rings with diameters of 60, 80, 100, 120, 140, 160, 180 mm have been
inscribed. A follow tube 60 mm in diameter by 80 mm long is placed on
the centre of the plate, filled with the slurry and trowelled level.
It is then removed vertically from the plate and the number of rings
covered by the slurry is noted. This test is similar to the concrete
slump test and has nearly as many detractors. However, it is a simple
test and if used regularly an acceptable consistency range of the
slurry will be quickly established. Test results outside of the
acceptable range will require an immediate investigation.

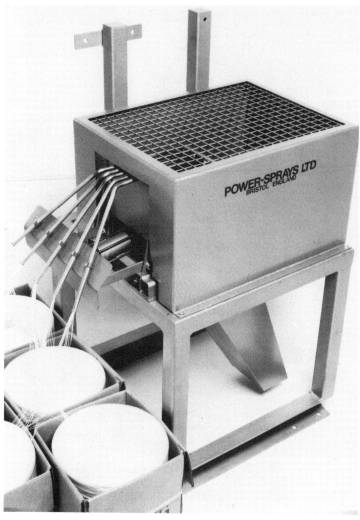

Fig. 4. A multi-roving glass fibre chopper.

5.2 Washout test

This is another simple test which is used in both the spray-up and
the premix processes to test the percentage glass fibre content and
the distribution of the fibres through the mix.

Samples of the fresh mix can be either cut from cast sample boards
or taken from the mixer during discharge. They are weighed in a
pre-weighed wire basket and the sand and cement are then removed by

washing out. The glass fibres remaining in the basket are oven dried and after weighing the percentage fibre content can be calculated.

The tests mentioned above help to control the quality of the GRC mix and are augmented by flexural tests and density tests performed on cured samples at the ages of 7 and 28 days (BSI 1984).

6 Conclusions

The addition of the glass fibres combined with a matrixof a high cement content but without aggregate above 1mm has necessitated the development of specialized mixers. As with all concretes the need for homogeneous mixes and uniformity of the workability has been recognized and simple tests have been developed to monitor them.

7 References

The Glass Fibre Reinforced Cement Association (1986), **Methods of Testing Glass Fibre Reinforced Cement (GRC) Material**.

The British Standards Institution (1984), **BS 6432:1984; Methods for Determining Properties of Glass Fibre Reinforced Cement**.

10 MIXING PROCEDURE OF FIBRE CONCRETE

J.-D. WÖRNER and H. TECHEN
Institut für Massivbau, Technische Hochschule Darmstadt,
Germany

Abstract
The processing of synthetic short staple fibres from the
delivery in containers till to the incorporation into concrete needs a special procedure. Because of their low density synthetic fibres are not easy to handle and to batch.
The fibres are usually pressed together in containers which
complicates the loosening and the separation. To avoid
bunches the matrix composition, the mixing sequences and
the mixer types may be changed.
Keywords: Delivery, loosening, batching, separation, incorporation of synthetic fibres in concrete

1 Introduction

It is well known that incorporation of short staple fibres
into any matrix requires particular care to ensure homogeneous fibre distribution.

The final product will exhibit weaknesses, if the fibres
are not distributed homogeneously. An inadequate flexural
strength, compressive strength, lower ductility, an increased shrinkage cracking tendency as well as unsatisfactory processing properties may be the result.

Polyacrylonitrile staple fibres tend to form bunches
like all synthetic fibres. These bunches must be completely
opened up and seperated into individual fibres, which must
be distributed homogenously in the matrix.

The application of short polyacrylonitrile staple fibres
is extremly wide. General application rules for deciding
which fibre incorporation method or which fibre types,
lengths, diameters and amounts are suggestive do not exist.
The decision depends on numerous factors that vary from
user to user and must therefore be determined in each
individual case.

- Type of mixer and production method,
- matrix components,
- matrix condition (dry-moist-wet) and
- finished product and its requirements profile

Special Concretes: Workability and Mixing. Edited by Peter J. M. Bartos. © RILEM.
Published by E & FN Spon, 2–6 Boundary Row, London SE1 8HN, 0 419 18870 3.

are the main important factors for a good workability and mixing.

In deciding on specific fibre types, lengths, diameters and amounts numerous different data have to be taken in consideration.

The density is an essential factor in comparing different fibre types because the volume content has a major bearing on mechanical properties as shown in table 1.

Table 1. Example of the relation between fibre material, density, weight and content

Fibre material	Density $[g/cm^3]$	Fibre weight $[kg/m^3]$ with fibre content of 1 % by vol.	% by vol. with fibre content of 10 kg/m^3
Acrylic	1,18	11,8	0,85
Polypropylene	0,90-0,98	9,4	1,06
Aramid	1,38-1,45	14,1	0,71
Carbon	1,70-1,90	18,0	0,56
Glass	2,45-2,68	25,7	0,39
Steel	7,85	78,5	0,13

2 Incorporation process of synthetic fibres

2.1 Principle
Fundamental trials and trials under service conditions led to the development of a method for the homogeneous incorporation of the staple fibres in moist and dry matrices. The individual stages in the process are:

- Delivery of containers
- Opening and loosening
- Gravimetric or volumetric batching
- Separation of the fibres
- Incorporation

2.2 From the container to batching
The fibres must first be loosened from the pressed bales and separated roughly. Machines for this purpose have already been tested.

Depending on the mixing process the fibres must be batched, a distinction being drawn between continuous and discontinuous operation. Because of the low and variable bulk density of the fibres gravimetric batching is generally preferable to volumetric batching.

2.3 Separating and incorporating
How well the fibres are incorporated depends greatly on the mixer and its operation, the type of fibre and the matrix composition.

2.3.1 Simultaneous separation and incorporation
The aim is to add the fibres unseparated to the mixer. The mixing system must be capable of separating the fibre bunches and distributing the fibres homogeneously in the matrix in one operation. Therefore some requirements of the mixing system are necessary.

The matrix must be whirled upwards or loosened so as to create room for incorporation of longer fibres.

The mixing elements must move the matrix on the counterflow principle so that the matrix components have a combing effect on each other, thus opening up the fibre bunches.

In the case of longer fibres additional high-speed mixing elements (for example whirlers) assist fibre separation by higher shear.

The separation and incorporation of the fibres must be effected within a short time. Lengthy mixing times may cause demixing and renewed compaction of the fibres.

Zones reached by mixing elements for only part of mixing time or dead zones where the fibres can accumulate are not satisfactory. Motion that causes uncoiling or bunching of the matrix has an adverse effect on the separation and incorporation of the fibres.

The requirements are met for example by the mixers shown in figs. 1 to 3.
The mixing systems mentioned were tested successfully in trials and in service conditions with various matrices such as slightly moist sand, concrete (maximum parcticle size 16 mm) and dry mortar. In all cases the different types of polyacrylonitrile fibres were added unseparated.

The mixing result depends on:
- the composition of the matrix (aggregate shape and size), distribution, cement content, moisture content and
- the type and amount of fibres used.

2.3.2 Sequential separation and incorporation
The commercial forced-circulation mixers used in the construction industry (also whirlers or satellite/planetary mixers) often have limited suitability or necessitate special methods of application for incorporating fibres in construction materials. The mixers operate in the shear range, loosen the matrix inadequately and do not produce a counterflow sometimes.

HOECHST developed for the polyacrylonitrile fibre DOLANIT a fibre separation machine, which can be used already commercially and can be adapted to any mixer. Fibre separation is effected by a high-speed brush roller, which

Fluidizing mixing zone

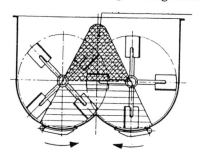

Fig.1. Double-shaft mixer
with cutter shaft

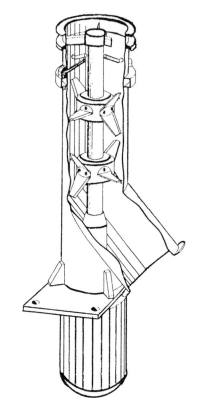

Fig.2. Flow mixer

 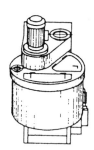

Fig.3. Pan mixer

loosens the individual fibres from the fibre bunches and ejects them into the moving matrix.

3 General information concerning incorporation of fibres

Fibres can be incorporated easier and in greater amounts in matrices that consist solely or largely of fine components. If the matrix is composed of coarse and fine material, it is advantageous to incorporate the fine components first and then add the coarse component.

- It is easier to incorporate the fibres into moist matrices than into dry matrices.
- The general rule for all matrices is that short fibres are easier to process than long ones.
- Thick fibres are easier to distribute than fine ones.

Generally applicable recommendations for a complete process from the fibre container to the finished product can be given only with reservations because of the wide variety of uses. The conditions pertaining in a particular case must be checked.

4 Practical suggestions for better processing and distribution of synthetic fibres

Variation of the mixing sequences
From practical experience the following procedures were positively judged:

a) ▪ mixing the concrete as usual and
    ▪ adding of fibres afterwards

b) ▪ mixing of fine aggregates, cement and water
    ▪ adding of fibres
    ▪ adding of coarse aggregates

c) as a) and using special separation equipment

d) as b) and using special separation equipment

e) ▪ drymix cement + fibres = premix
    ▪ premix + ordinary mixing procedure

These five possibilities show different effectivity with regard to the possible fibre amount and fibre geometry. Fig. 4 shows the applicability of the different mixing procedures for DOLANIT 104/6.

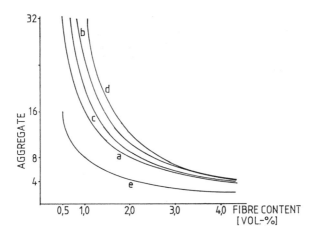

Fig.4. Limit curves for applicability of the different
mixing procedures for DOLANIT 104/6

Varying the binder content
- Increase the cement content or
- Add fly ash

Use of plasticizer or super plasticizer
The water absorption of polyacrylonitrile fibres is
negligible but the large surface of the fibres binds
moisture by adsorption. In order not to increase the water
content unnecessarily and thereby causing loss of strength,
the use of plasticizers or super plasticizers is recommended. Fig. 5 shows the fibre content as a function of the
water/cement ratio. This is based on the following facts:
  Aggregate 0/8 mm
  Cement PZ 45 F (high early strength) content of 400 kg/m^3
  Polyacrylonitrile fibre DOLANIT, d = 104 μm, l = 6 mm
  Super plasticizer 2 % by mass of cement

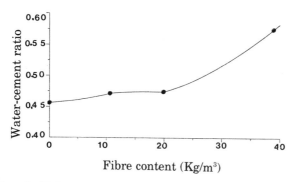

Fig. 5. Relationship between water/cement ratio and workability

Table 3 lists some recipes that have been tested in practice. They meet the requirements for miscibility and material properties achieved. Mixers commonly used in the industry were employed, some in combination with fibre separation machines or process variants.

Table 3. Examples for mix proportions for mortar and concrete which have been tested in practice and meet the requirements for miscibility and for material properties

Use	Aggregate mm	Cement kg/m^3	Filler kg/m^3	Water kg/m^3	Admixture %	Fibre	Content kg/m^3
Industrial flooring	0/8	260	--	150	3,4 FM	DOLANIT 104/6	5
Wall and flooring flags	0/32	320	60	190	2,0 FM	DOLANIT 104/12	10
Catching basins	0/22	360	--	180	0,5 FM	DOLANIT 104/6	10
Flooring flags	0/16	280	120	185	1,5 FM	DOLANIT 104/6	10
Bridge caps	0/16	350	--	170	0,5 FM 0,2 LP	DOLANIT 104/6	10
Facade elements	0/16	430	--	200	1,0 FM	DOLANIT 104/12	12
Bullet-proof facade elements	0/16	420	--	175	1,0 FM	DOLANIT 104/6	17
Noise screens	0/8	420	--	170	2,0 FM 0,6 LP	DOLANIT 104/6	15
Water channels	0/4	500	--	175	2,0 FM	DOLANIT 104/6	50

FM = super plasticizer
LP = air-entraining agent

11 MIX DESIGN APPROACH FOR FIBRE REINFORCED MORTARS BASED ON WORKABILITY PARAMETERS

G. PEIFFER and P. SOUKATCHOFF
Centre de Recherches P.A.M., Pont-à-Mousson, France

Abstract
This paper presents a part of a research programme which aims to develop a general method of mix design of fibre reinforced composites and a method of design of manufactured elements based on identified stress-strain relations and modelling. The paper deals with the formulation of mortars and the mechanical properties of the optimised composite materials.
Keywords: Fibre Reinforced Materials, Slump Test, Mix Design, Workability, Steel Fibres, Mix Additives

1 Formulation of 3 different FIBRAFLEX prototype mortars

The fibres considered in this study are FIBRAFLEX amorphous metallic fibres of 30 mm length produced by SEVA, France.
The method is based on two steps:

Step 1   Formulation of slurries with cement, pozzolana, water, superplasticizer. The content of silica fume or metakaolinite is 10% by weight of the cement. The amount of superplasticizer is optimised with respect to rheological behaviour of the slurry. Different water/cement ratios are investigated for the 3 types of slurries: cement, cement + condensed silica fume, cement + metakaolinite (MK E1), in order to get the same "flowability" of each slurry.

Step 2   For each optimised slurry and for a content of FIBRAFLEX of 1% by volume the workability of mortars is measured by a slump test as a function of the slurry volume content in the mortar.

A reference slump value was fixed throughout the whole testing programme to obtain the right placing behaviour for all the mortars. The results obtained at different

Special Concretes: Workability and Mixing. Edited by Peter J. M. Bartos. © RILEM.
Published by E & FN Spon, 2–6 Boundary Row, London SE1 8HN, 0 419 18870 3.

stages of the study are described in the following
sections.

1.1 Optimisation of the content of the superplasticizer (SP) for each slurry

Dependance of the flow time (in funnel test) on the
content of the plasticizer is shown in Fig. 1. At this
stage, the water/cement ratio was the same for the 3
slurries (w/c = 0.35). Each curve has two typical slopes.
For small amount of plasticizer the flow times vary
rapidly, then after a threshold value of SP, which depends
on the type of the slurry, the flow time remains in a very
small range of variation.

The optimal amount of SP was chosen as a value which
was slightly higher than the very beginning of the plateau
of the curve. Possible variations of the quality of the
industrial products must have been taken into account so
our choice was on the safe side considering further
applications on site. The content of SP is given as a
percentage of the weight of the cement in Tab. 1.

Table 1. The content of SP in the slurry

Type of slurry	Optimal content of SP
Cement	1.5%
Cement + silica fume	3%
Cement + metakaolinite E1	3%

These values were fixed for the further steps of the
formulation (either mortar or concrete). It was thought
that the optimal content of SP would not vary
significantly when the w/c ratios were changed. It was
also believed that sand or coarse aggregates would have no
influence on the maximum efficient content of SP.

1.2 Influence of the w/c ratio on the flow time of the slurries

The rheological behaviour of the three types of slurries
for different values of the w/c ratio is investigated in
this section. The content of SP was set to the optimum
value for each slurry. The results of the flow time
measurements are plotted in Fig. 2.

As a general rule, the flow time increases for each
slurry as the w/c ratio decreases, but we noticed a fluid
consistency of the slurries for w/c values as low as 0.25.
Comparison of the curves shows that for a given
consistency, the addition of 10% metakaolinite by weight
of the cement requires a w/c ratio higher than for a 10%
CSF and than for a plain cement. The differences between
the three curves seem to vanish when w/c ratio reaches the

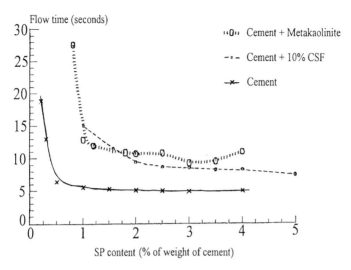

Fig.1. Effect of superplasticizer at w/c = 0.35
on the flow behaviour of slurries

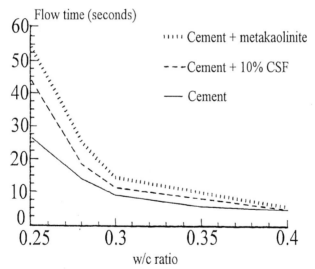

Fig.2. Effect of the w/c ratio on the flow
time of slurries

value of 0.4.

This study permits us to formulate slurries with a constant flow time which will be chosen at the next step.

1.3 Workability of V_f= 1% FIBRAFLEX reinforced mortars as a function of the volume of the slurry

The aim was to find the right volume of slurry per m^3 of a mortar which was necessary for obtaining a given consistency of mortar measured by a slump test. Extensive work has been done in our laboratory to measure the variations of slump values of mortars versus volume of slurry per m^3 for each type of pozzolanas. Typical curves are shown in Fig. 3. For a given type of pozzolana, higher flow time of the slurry demands a higher volume of the slurry to achieve the given workability. When pozzolanic additives are used in mortars, this demand decreases comparing to a plain slurry.

For example, for a slump of 15 cm and a constant flow time of slurry of 10 s, the volume of the slurry per m^3 of the mortar is 20 - 25 litres lower when a CSF cement slurry is used rather than a plain cement slurry.

1.4 Mortars of a constant slurry flow time and a constant slump

The above study was necessary in order to choose a correct flow time of the slurry and a reference workability of the mortar. One should keep in mind that the aim of the method is the mix design of mortars with different FIBRAFLEX fibre contents, ranging between 0 and 1.6 % by volume, and concretes with different types and contents of fibres. All these composites should be based on the same slurry formula in order to keep the physical and chemical conditions of the paste and the fibre-paste interface as constant as possible. The flow time chosen for the different slurries should allow for a good workability to be reached without segregation of the slurry within the whole range of studied parameters (content of fibres, type of aggregate, type of fibre, etc.) Some preliminary placing under vibration tests showed that we could set the flow time to 7 seconds and the slump value to 10 cm. The formulas of the slurries are shown in Table 2.

Table 2. Formulas of the slurries

Type of slurry	Superplasticizer	w/c ratio
Cement	1.5%	0.3
Cement + silica fume (10% by weight)	3%	0.37
Cement + metakaolinite E1 (10% by weight)	3%	0.385

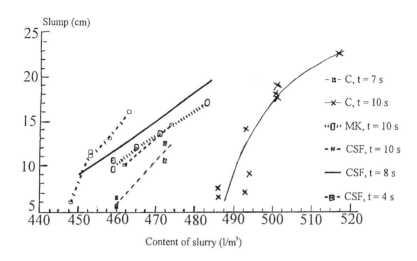

Fig.3. Effect of the type and flow time of slurry on
the workability - volume of slurry relation

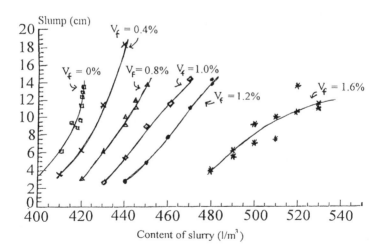

Fig.3a Effect of the slurry content on the
workability of FIBRAFLEX mortars

The compositions of the three prototype mortars are given in Table 3. FIBRAFLEX fibre content is 1% by volume.

Table 3. Composition of the prototype mortars

Type of slurry	Volume of slurry (l/m^3)	Slump
Cement	472	10.5 cm
Cement + silica fume (10% by weight)	457	9.5 cm
Cement + metakaolinite E1 (10% by weight)	460	10.0 cm

2 Mechanical characteristics of three types of mortars

For each type of mortar, 40 kg batches were mixed and then plates with dimensions 1000 x 650 x 20 mm were cast. The measured slump values of the fresh mixes ranged between 11.0 and 11.5 cm.

The following curing regime was adopted: 1 day in covered moulds and 27 days at 20°C and 50 % RH.

14 samples were sawn out from the plates at the age of 14 days. The dimensions of the samples were 350 x 80 x 20 mm. Four point bending tests were carried out using a 2500 kN INSTRON testing machine at constant cross-head speed of 2mm per minute. The span was 300 mm. The AFNOR porosity was also measured. The test results can be found in Tab.4.

Table 4. Test results

	Cement	Cement + CSF	Cement + SP
LOP (MPa)	8.5	10.9	10.4
MOR (MPa)	12.3	14.9	13.2
EPS (%)	0.22	0.22	0.17
E modulus (MPa)	30 000	33 000	40 000
Porosity	15%	10.6%	13.9%

After these results were obtained, it was decided to continue the study with the optimised cement + CSF slurry.

3 Formulation of FIBRAFLEX reinforced mortars with V_f from 0% to 1.6%

At the previous stages the content of fibres was kept at 1% by volume. This allowed us to change the parameters of

1% by volume. This allowed us to change the parameters of the binder slurry only and we finally chose the optimised cement + 10% CSF slurry. The next aim was to formulate fibre reinforced mortars based on the same slurry, with different fibre contents: 0%, 0.4%, 0.8%, 1.0%, 1.2% and 1.6% by volume. The slump value was required to be equal to that of the finally chosen formulas, ie. 11.5 cm.

The method described in Section 1.3 was used. For each fibre content several batches with different slurry contents per volume were prepared. The total mixing time was 13 minutes and the volume was measured within 3 minutes. Three measurements of slump were carried out for almost every batch. In general, the slump value increased as the slurry content increased for a fixed fibre content. For a fixed slurry content, the workability decreased with increasing fibre content. Therefore, the slurry content which is necessary to achieve a fixed workability should be higher when using higher fibre contents.

The slump values vs. slurry contents are plotted in Fig. 3a for each of the 6 fibre contents. As can be seen, the rheological behaviour of the 1.6% mortar appears to be very different from that of the lower fibre content mortars. Beyond the values of 9 to 10 cm the slump increases at a very low rate when increasing the slurry content. This certainly shows that for a fibre content as high as 1.6% there is no flow possible due to gravity only. One may assume that the fibres in this case form a very tight mesh which is quite stable by itself, increasing the volume of the slurry or using a more fluid slurry would certainly cause segregation. However, we noticed that a reliable slump measurement was difficult to perform because of great differences in the behaviour of the mortar. On the contrary, for the lower fibre contents, the mortar seemed to behave like a suspension for not connected fibres.

With regard to the future industrial process of manufacturing precast elements some tests were performed measuring workability of mortars versus total duration time of mixing. The measurements of slump were carried out every two minutes on mortars with identical initial workability (ie. roughly 11 cm slump after 13 minutes of initial mixing). After each measurement of slump the mortar was poured back into the batch and mixed again for two more minutes. The results are shown in Fig. 3b.

Workability increased with the time of mixing, the slope of the curves being sharper for standard and low fibre content mortars. It appears that the high fibre content lowers the effect of variations of the flow properties of mortars. These results led to the choice of mortar compositions with 11 cm slump. For each fibre content the composition of the mortar is defined by the cement + CSF slurry content as shown in Table 5.

Table 5. Composition of mortars

Fibre content per volume	0.0%	0.4%	0.8%	1.0%	1.2%	1.6%
Volume of slurry (1/m^3)	419	430	445	460	470	530

4   Manufacture of plates with different volumes of FIBRAFLEX in optimised mortars

For dimensions of the plates see Section 2, the formulas of mortars are described in Section 3. The manufacturing process was as follows: The fresh mortar was poured into a mould in 3 layers under vibration and each layer was rolled. For each type of the composition, 2 plates were manufactured. Slump values were measured for each batch, the results ranged from 7 to 8.5 cm. These results appear to be significantly different from the previous work. The reason is probably variation of raw materials especially in sand grading.

Nevertheless, it was decided not to change the compositions because the value of the slump obtained (8 cm) was still correct enough for placing mortars in moulds under vibration. The only exception was the formula with V_f = 1.6%, where the slurry content was changed from 530 to 500 1/m^3 to avoid segregation problem.

The flexural strengths of the different materials are given in Table 6.

Table 6. Flexural strength of the mortars

V_f (%)	0.0	0.4	0.8	1.0	1.2	1.6
MOR (MPa)	8.2	7.9	10.1	12.6	12.2	13.4

The results of measurements of porosity of different mortars are given in Table 7.

Table 7. Porosity of the mortars

V_f (%)	0.0	0.4	0.8	1.0	1.2	1.6
Porosity (%)	8.2	7.9	10.1	12.6	12.2	13.4

As can be seen from the Table 7, the values of porosity are very low in comparison with common mortars, thanks to the previously described optimisation method.

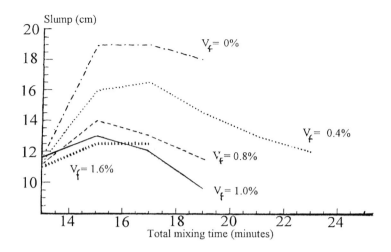

Fig. 3b Variation of the workability with mixing
time (mortars of the same initial workability

5 Conclusion

A general experimental approach to the mix design of
premix-fibre reinforced composites was tested. It is based
on a description of the material which consists of two
components, a cementitious slurry and a granular skeleton
(sand, gravel and fibres). The method allows further
modifications of the binding slurry; only little more
experimental work is to be done on fresh mortar or
concrete.

The mortar compositions determined by the previous
laboratory study can be easily mixed and placed in the
mould using the usual equipment of the manufacturing
plant.

In-plant manufacturing tests, which included varying
both the thickness of the slabs (up to 30 cm) and their
design, confirmed the feasibility of the slabs in a wide
range of dimensions of cross-section.

6 Acknowledgement

This work formed a part of the European BRITE programme
which ended in 1992.

12 THE EFFECTS OF POLYPROPYLENE FIBRES IN FRESH CONCRETE

G. McWHANNELL
Fibermesh Europe, Chesterfield, UK

ABSTRACT
The use of polypropylene fibres in concrete for effective
plastic shrinkage crack control purposes, and to improve
the hardened surface properties of concrete, has been
established over a number of years and is based on a wide
range of test data, most of which has been produced by the
Fibermesh Company. In this paper it is proposed to look at
the production aspect of polypropylene enhanced concrete,
the effects the presence of these fibres have on the rheology of the concrete and how it can be checked and tested.
Keywords: Polypropylene fibres, Fibrillated, Monofilament, Thixotropic, Plastic crack control.

1 Introduction

The concept of adding fibrous material to a cementitious
mix is not new - in Biblical times straw was added to brick
clay, and not so long ago it was common practice to add
horse hair to plaster and render for crack control purposes.

So what happened to the concept - it lay dormant for a
while, in fact it was not until the 1960's when S Goldfein
revived and developed the idea using the new synthetic
fibres.

Various types of material were considered - Nylon, Poly-
ester, Polyethylene, Polypropylene etc, but it was soon
shown that the best material to put in to a wet, highly
alkaline cementitious environment was Polypropylene.

It was the early 80's before the fibre was developed
such that it could be effectively handled by a truck mixer,
gave no trouble in concrete placing and became economical
enough for serious consideration in routine concrete production.

Much of this early development is largely attributed to
Jes Cutlip, formerly of Delta Concrete Co, and Wayne Freed
of Synthetic Industries Inc - a major producer of

Special Concretes: Workability and Mixing. Edited by Peter J. M. Bartos. © RILEM.
Published by E & FN Spon, 2–6 Boundary Row, London SE1 8HN, 0 419 18870 3.

polypropylene yarns, and in 1983 the Fibermesh Company was formed as a subsidiary of Synthetic Industries Inc, USA.

In 1989 Fibermesh opened their UK production facility, to manufacture polypropylene fibres to BS5750/ISO9002 Quality Assured Standards and British Board of Agrément (BBA) performance assessment.

2   Polypropylene Fibres

Fibermesh polypropylene fibres have been used very success-fully in millions of square metres of ground supported floor slab throughout the UK, Europe and the rest of the World, to control plastic shrinkage cracking and enhance the hardened surface properties of the concrete, without affecting the structural integrity of the concrete. In fresh concrete Polypropylene fibres will also reduce bleed-ing and minimise settlement; but, questions still arise concerning fibre type, fibre length, fibre dosage, fibre addition and mixing, workability, and testing of the fresh fibre enhanced concrete.

2.1   Fibre Type
There are commonly two types of polypropylene fibre man-ufactured for use in concrete, namely a) Fibrillated - a fibre of near rectangular cross-section and irregular surface texture, formed when the mixing action of the concrete opens the fibrillated fibre bundles and separates the fibres; and b) Monofilament fibre - a fibre of circular cross section, smooth surface texture and generally finer than the fibrillated fibre.

It is generally accepted that due to the irregular sur-face texture and hence the improved mechanical bond, the fibrillated fibres give significant improvements to the hardened properties of the concrete and are therefore best used in industrial and warehouse flooring, as well as external paving and hard-standings. The Monofilament fibres on the other hand show significant benefits in the lighter duty slabs especially where fine and particularly hand finishing is required.

2.2   Fibre Length
In the 1980's when fibres were first introduced, the fibre lengths were in the range of 38mm to 50mm with some vague reference to bond length. However a number of difficulties were encountered particularly when trying to pump fibre enhanced concrete. It was found that the grill over the concrete pump hopper tended to filter out the fibres as these long fibres wrapped themselves around the bars. A succession of length reductions were tried and the optimum general purpose length of the fibrillated fibres was found to be in the order of 19mm, and 12mm for the Monofilament fibres. This does not mean to say that other lengths are

not or cannot be used. The main side effect of these fibre length reductions was the reduction of the thixotropic effect experienced when using fibre enhanced concrete. To some people, such as the precast pipe manufacturers, this is a disadvantage, therefore a supply of 38mm and 50mm fibres is still available to this industry for their special requirements.

2.3   Fibre dosage
Probably the most critical aspect of fibre enhanced concrete and the most widely researched. It was through this research that in 1983 Fibermesh established the 'Construction Norm' of 0.1% by volume or what is more commonly known as 0.9kg or 900 gms/cubic metre of concrete. Tests have shown that increasing the dosage will have little effect on the majority of plastic and hardened properties, whereas reducing the dosage can have a marked and detrimental effect on all the properties of the concrete regardless of fibre length or diameter. Many arguments have been put forward as to why this should be, but upto now no one has been able to give a technically satisfactory answer, backed up by test data.

2.4   Fibre Addition and Mixing
Initially it was considered necessary to add the fibres gradually into the mix, in a manner not dissimilar to that used for steel fibre; but testing and experience has shown that dependent on the mixer type, mixing time can range from less than one minute in a high sheer forced action pan mixer to five minutes for a truck mixer, regardless of when it is added. This in turn has lead to the development of the Fibermesh Fas-Pak™ degradable bag which has been specially developed for truck mixed concrete. The bag itself is made from a fibrous inert amorphous carbohydrate polymer and has been tested and shown to have no effect on the concrete. In general fibres are packaged in 0.9kg bags, both plastic and degradable, but for the convenience of some concrete producers, other bag sizes are available.

2.5   Workability
As mentioned earlier under 2.2 fibre length, polypropylene fibres have a thixotropic effect on the fresh concrete mix. This can have major advantages and perceived disadvantages. It is common practice to believe that the concrete slump test has a direct relationship to workability - it may have with plain concrete, but it does not with fibre enhanced concrete. Tests have actually shown that although fibre enhanced concrete may have a slightly lower slump than an identical plain concrete, it will have a greater workability! In practical terms the thixotropic effect is shown as a resistance to slip when casting on gradients or slopes and is evident by the noticeable reduction in bleeding and

settlement, maintenance of level and flatness, and plastic shrinkage crack control.

2.6   Testing
How do you test fibre enhanced concrete for workability, fibre content, fibre distribution and plastic shrinkage crack control?

2.6.1   Workability
Variations in workability can be successfully measured by the Slump test, but to obtain a more accurate measurement, a Dynamic test is required (to overcome the thixotropic effect). For this purpose, the use of a simple flow table test, would be a practical site/field test, or the more accurate Compaction Factor or V-B test if laboratory facilities were available or near to hand.

As mentioned before, accurate comparative laboratory tests have shown a small reduction in the slump of fibre enhanced concrete compared to an identical non-fibre concrete, but the same concrete also showed a small increase in workability as measured by the Compaction Factor test.

2.6.2   Fibre Content
Fibre content can be checked, easily at the batch plant by observing the addition of the fibre; fairly easily in the mixed plastic state, by performing a washout test on a known volume of concrete; and by a somewhat expensive acid washout test on a sample of hardened concrete. In both the washout tests, more than one sample would be required to obtain an average result.

Most major concrete producers in the UK are members of the Quality Scheme for Ready Mixed Concrete (QSRMC) and are required to perform internal audits on materials used, compared to concrete produced etc, thus the onus is very much on the concrete producer to ensure that the correct dosage is used at all times.

2.6.3   Fibre Distribution
How can you tell whether the fibres have been distributed uniformly and in a multi-directional manner throughout the concrete mix? The simplest way is to examine the Fresh concrete. If Fibrillated fibres have been used, the 'individual' fibres should be easily visible either as the concrete is discharged from the mixer/mixer truck, or, in the freshly laid concrete. If the fibres still appear to be joined together, then mixing has not been completed. Fibrillated fibres have a type of built in quality control mechanism!

Due to the fineness of the Monofilament type of fibre, it is much more difficult to 'see' the fibres, and therefore correspondingly more difficult to determine fibre distribution. However, by taking a sample of concrete and gently pulling it apart and examining the exposed faces, many hundreds of individual fibres should be seen.

In both cases fibre distribution can ultimately be determined, if required, by the microscopic examination of 'thin sections'; but this is an expensive test which need not be necessary if the concrete is mixed properly in the first instance.

2.6.4   Plastic Shrinkage Crack Control

The major benefit of polypropylene fibres in concrete and possibly the most difficult to test. After considerable deliberation, the British Board of Agrément (BBA), have approved the use of the Modified Kraai Slab Test (see Appendix A) to determine the reduction in crack area, and the FCB Trondheim Ring Test to determine the reduction in crack width. Unfortunately both tests require very accurate monitoring and can only be carried out under stringent laboratory conditions.

After the concrete has been laid, compacted and finished in accordance with the specification, it must be cured. Polypropylene fibres are not a substitute for curing agents/compounds, although the presence of the fibre does reduce the rate of moisture loss. There is no substitute for good concreting practice.

3   Good Concrete Practice

Much has been written and even more has been said about good concrete practice; but equally, much tends to be forgotten when it comes to 'getting the job done'. It is therefore important to remember that adding extra water to improve the concrete 'workability' is not to be recommended. If a more workable concrete is required then it should be designed as such. Most of the national and many of the local concrete producers in the UK have used Fibermesh polypropylene fibres and are happy to give a fibre mix design to suite purpose. In order to ensure proper distribution and separation of the fibres in truck mixed concrete, good practice not only requires the fibres to be mixed for five minutes before leaving the plant, but strongly recommends that the concrete should be remixed for a minute or two immediately prior to discharge.

4   Conclusion

The addition of polypropylene fibres to a concrete mix is a simple and cost effective method of obtaining a high degree of plastic shrinkage crack control and, reduced bleeding and settlement, without affecting the actual workability or placability of the concrete.

Where required, simple field procedures are available to

check and/or test for workability, fibre content and fibre distribution; but as with all concrete, for the best results it must be backed up by good concrete practice.

5   References

American Concrete Institute, (1988) ACI 544.2R-88 – 'Measurement of Properties of Fibre Reinforced Concrete'.

British Board of Agrement (BBA), (1992) Watford, England, – 'Polypropylene Fibres for Concrete – Test Programme'.

British Standards Institute, (1979) BS5750' Quality Systems'.

Concrete Society Technical Report, (1992) No.22 'Non-Structural Cracks in Concrete.

Dhal. P.A, (1985) 'Plastic Shrinkage and Cracking Tendancy of Mortar and Concrete containing Fibermesh'. FCB Cement and Concrete Research Institute, Trondheim, Norway.

Gold, S., (1990) 'Investigation of Fibermesh', A.C.T. Project, Ready Mix Concrete Ltd, London, 1990.

Goldfein. S., (1963) 'Plastic Fibrous Reinforcement of Portland Cement'. Technical Report No. 1757-TR, US Army Engineering Research & Development Laboratories, Fort Belvoir VA.

Kraai P.P, (1985) 'A proposed test to determine the cracking potential due to dring shrinkage of concrete'. Concrete Construction Publication, Addison, Illinois.

Soroushian.P., Khan.A., Hsu.J-W., (1992) 'Mechanical Properties of Concrete Materials Reinforced with Polypropylene or Polyethylene Fibres'. ACI Materials Journal, Vol 89, No.6.

Vondran.G., 'Making More Durable Concrete with Polymeric Fibres', ACI SP100-23 Concrete Durability pp 377-396

6   APPENDIX A.   THE MODIFIED KRAAI SLAB TEST

RESISTANCE TO PLASTIC CRACKING OF CONCRETE SLABS
British Board of Agremént, Watford, England and Fibermesh Europe Ltd, Chesterfield, England

SPECIFICATION
SCOPE
This specification describes an effective method of determining and/or comparing the resistance of various types of concretes, including fibre enhanced concrete, to the formation of plastic shrinkage cracking. The basis of the test is to create an environment which will give the exposed surface of the concrete an evaporation rate in excess of

the critical 1 Kg/m^2/hr for the formation of plastic shrinkage cracks. In order to emulate normal slab conditions, the concrete should be subjected to edge restraint.

1 METHOD

1.1 Apparatus
Prepare formwork for 3 No. slabs each 600 mm x 600 mm x 50 mm deep in a wind tunnel arrangement, such that the exposed surface of the samples will be subjected to an evaporation rate in the order of 1.25 Kg/m^2/hr.

Formwork should comprise of timber sides and Melamine/Formica coated base, and should be rigidly constructed with the joints sealed to prevent loss of water and fines.

Firmly attach two perimeter restraint bars of diameter 5mm to the internal sides of the formwork at mid height in each of the slabs. The bars should be approximately 25 mm in from the outside edge and 25 mm apart.

Provide 4 No. 100 mm polystyrene cube moulds for each test run.

Provide copies of current Certificates of Calibration relating to equipment for the measurements of temperature, humidity and wind speeds.

1.2 Test Specimens
Prepare at least three slabs of size 600 mm x 600 mm x 50 mm from each concrete to be tested. See Note 3.1.

The slabs should be cast on a lightly oiled melamine/formica surface. Sides of the formwork should not be oiled.

Fully compact the concrete (method of compaction and duration to be recorded) and screed and steel float, using the same number of passes to reduce the variation in finishing each slab. See Note 3.2.

In addition, for each set of slabs prepare 4 No. 100 mm cubes.

Photograph the prepared specimens and comment on the quality and nature of finish achieved and degree of fibres exposed on the surface.

DO NOT CURE SLABS.

1.3 Procedure
Immediately after trowelling condition the slabs for 24 hours as below.

Subject the samples to an evenly distributed air stream of temperature 28°C (± 2°C) relative humidity 40% (± 5%) and wind speed of 25 km/h to achieve a rate of evaporation of the order of 1.25 kg/m$_2$ per hour, See Note 3.3.

After this, condition the slabs for 28 days in laboratory controlled conditions at 23°± 2°C and relative humidity 50% ± 5%.

2 Results

Continuosly measure and record the temperature, relative humidity and wind speed of the air throughout the testing period.

Measure and record the weight loss after ½ hour, 1 hour, 2 hours, 3 hours, 4 hours, 5 hours, 6 hours (and then at 24 hours) of the cubes at the outlet of the wind channel.

Record the time of the first crack.

Measure and record the crack development (length and width) on the surface of the samples.

Photograph crack development after each of the above time intervals.

Calculate the crack area for each slab from the following:

Crack area = length of crack $\left(\dfrac{\text{Max width} \pm \text{Min width}}{2}\right)$.

After 28 days wet the surface of the slabs and photograph the crack pattern. Turn each of the slabs over, photograph and record the number and extent of cracks where water has penetrated to the underside.

Compare the length of the cracks on the bottom with corresponding cracks on the top surface.

3 General Notes

3.1 Where comparison tests are being carried out, ie Control: Fibres, the concrete mix design used must be the same.

3.2 As the tests are mainly concerned with the quality and nature of the surface of concrete it is important that consideration be given to the method of finishing the concrete.

3.3 An increase in temperature and/or a reduction in relative humidity is permissible provided the rate of evaporation is of the order 1.25 kg/m² per hour.

It may be necessary to increase the temperature of the aggregates and water for use in the mix. Temperatures must be recorded.

It may be necessary to conduct a number of trials to establish that the rate of evaporation required can be consistently achieved throughout the tests.

13 COMPARATIVE MEASURES OF WORKABILITY OF FIBRE-REINFORCED CONCRETE USING SLUMP, V-B AND INVERTED CONE TESTS

C. D. JOHNSTON
Civil Engineering Department, University of Calgary,
Calgary, Canada

Abstract
Methods of measuring the workability of fibre-reinforced concrete are evaluated. Measurements of slump, V-B time and inverted cone time are compared for a range of mixtures and their sensitivity to change in fibre and mixture parameters established. Fibre variables include amount, length, aspect ratio and type, with the bulk of the results obtained using steel fibres and the remainder using alkali-resistance glass strand or fibrillated polypropylene. Mixture variables include the volume fractions of cement paste and coarse aggregate. The V-B and inverted cone tests are found superior to the slump test in terms of sensitivity to change in fibre and mixture parameters and their advantages and limitations are identified.
Keywords: Fibres, Concrete, Workability, Measurement Techniques, Precision.

1 Introduction

Fibre-reinforced concrete (FRC) is one of several special concretes that differs significantly from normal concrete with respect to measurement of workability and mixture proportioning. Fibre parameters such as type, length, aspect ratio and concentration in the concrete matrix have a profound effect on workability measured by any method. In this paper workability is evaluated in terms of slump, V-B time (B.S. 1881), and time of flow through inverted slump cone (ASTM C 995) subsequently termed inverted cone (I.C.) time.

Mixture proportions may also affect both the value of workability determined by any of these test methods and its variability. Since the fibres are accommodated solely within the cement paste, the consistency of that phase coupled with its relative volume fraction and the volume fractions of fine and coarse aggregate in the mixture may be expected to influence the value of workability obtained by any test method. Unsatisfactory mixture proportioning for the fibre type, amount and mixing process selected may fail to produce a reasonably uniform fibre distribution throughout the mixture causing the variability of workability measurements to be abnormally high. Therefore, properly selected fibre-mixture combinations and an effective mixing process are essential for meaningful assessment of any workability-measuring technique.

All three workability tests discussed are susceptible in varying degrees to operator-associated variables and subjective judgements of the test end-point, so the precision with which each workability measurement can be established depends

Special Concretes: Workability and Mixing. Edited by Peter J. M. Bartos. © RILEM.
Published by E & FN Spon, 2–6 Boundary Row, London SE1 8HN, 0 419 18870 3.

on the care and experience exercised by the operator. All the measurements reported herein were performed by inexperienced undergraduate students working in group laboratory projects that are part of the course requirements for a final year civil engineering materials course taught by the writer. Some learned the necessary skills more quickly than others and some exercised greater care than others, so the reliability of the results probably lies between what can be expected of meticulous trained research technicians working in a laboratory and the inexperienced sometimes careless operators encountered on many job sites.

To increase the time window within which workability measurements can be performed without being affected by the initial stage of the setting process, sugar was included in the mixing water to retard setting in most of the data reported.

2 Philosophy of Workability Tests for FRC

Since FRC is nearly always placed using vibration and fibres impart considerable cohesion to FRC mixtures under static conditions, the slump test performed in the absence of vibration on these abnormally cohesive mixtures is not a good meausre of workability and associated handleability and placeability, except perhaps for the highly fluid range of consistency that can sometimes be achieved using superplasticizing admixtures. A test that incorporates vibration is preferable. While the V-B test that has long been used outside N. America for many types of stiff concrete is one alternative, its cost and perceived unsuitability for site use are possible reasons for reluctance to accept it within N. America. Consequently, an alternative based largely on the cheapness, portability and ready availability on site of the necessary equipment (slump cone, internal vibrator and bucket) led to development of the inverted slump cone test and its recommendation for use with FRC by ACI Committee 544 (1978). Both the V-B and I.C. tests measure primarily the mobility of the freshly mixed FRC under vibration although the test sample in both cases must exhibit some degree of stability when placed in the test container prior to the start of the test. Once vibration commences flow or mobility is the primary phenomenon observed. However, both samples may undergo some consolidation and associated manifestation of compactability by the time the test end-point is reached. Both have limitations when the mixture is too stiff to flow under the test source vibration or so fluid that the test time becomes too short to be accurately determined. In the latter case, slump or DIN flow measurements may have merit, but the precision of such measurements is uncertain.

3 Workability Measurements for Steel FRC

These measurements are influenced by fibre content, aspect ratio and physical length.

3.1 Effect of Fibre Aspect Ratio
A previously published comparison (Johnston 1984) demonstrates that both the V-B and I.C. tests are effective in distinguishing the effects of fibre content and aspect ratio on workability (Fig. 1). Slump was less sensitive in distinguishing these effects as the slump range was only 13-44 mm, even though most of the mixtures were readily moved by vibration with I.C. times ranging between 9 and 30s and V-B times between 3 and 8s. This investigation was perhaps unique in

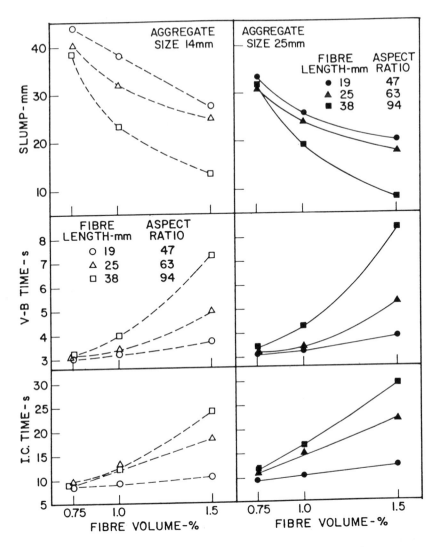

Fig.1. Effect of fibre aspect ratio on workability for
0.4 mm diameter steel fibres.

that three measurements were made by four different operators on each mixture,
thus establishing single-operator and between-operator variability of the results
(Table 1). Generally, the coefficients of variation are much lower for the V-B and
I.C. data than for slump. Relationships between results from the three different
tests were also established, the most interesting being the apparently linear
relationship passing close to the origin between V-B and I.C. times, I.C. time =
3.7 (V-B time) -1.2 (Fig. 2). This supports the view that both tests are measuring

Table 1. Summary of average values of between-operator and single-operator standard deviations and coefficients of variation.

| Type of Test | Between Operator | | Single-Operator | |
	Standard Deviation	Coefficient of Variation, %	Standard Deviation	Coefficient of Variation, %
Slump	6 mm	12-30[a]	5 mm	11-27[a]
V-B	0.32-0.49 s[b]	8.0-12.2[b]	0.25-0.36 s[b]	6.4-9.1[b]
Inverted Cone	0.84 s	6.2	0.55 s	4.2

[a]Significantly correlated with the mean. Increases with the decrease in slump for slumps of 13 mm or less.
[b]Limits shown represent 14- and 25-mm maximum size aggregate, respectively. Values appear to increase with aggregate maximum size.

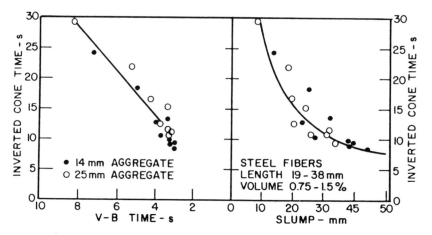

Fig.2. Relationships between I.C. time and V-B time or slump for 19, 25 and 38x0.4 mm steel fibres.

the same phenomenon on two different time scales, namely the placeability of FRC under vibration.

3.2 Effect of Fibre Length

In the previous investigation the longest fibre was 38 mm, relatively small compared with the 100 mm diameter orifice in the inverted cone. However, when the fibre length regardless of its aspect ratio becomes large relative to the cone orifice diameter, its physical length may interfere with flow of the FRC through the cone. Comparing relationships for FRC's containing 75 mm long crimped crescent-shaped steel fibres with those containing 25 mm long 0.4 mm diameter circular fibres (Fig. 3.), the I.C. time increases much more sharply for the longer fibre than would be expected from Fig. 1 simply from their difference in aspect ratio. For example, at a fibre content of 1% of 0.4 mm diameter wire fibre the I.C. time for the 25 mm (63 aspect ratio) is 13-15s in both Fig. 1 and Fig. 2 and is increased at most by 3s by increasing the aspect ratio to 94 in Fig. 1. In contrast, changing to the 75 mm (approximately 80 aspect ratio) fibre in Fig. 2 increases the I.C. time to more than 25 s, at least double the value for the 25 mm fibre. Clearly, the physical length of the fibre is important in that long fibres interfere more with flow of the FRC through the cone.

The physical length of the fibre seems also to be important in the V-B test because the V-B time in Fig. 2 increases much more sharply for the longer 75 mm fibre than would be expected simply on the basis of the difference in aspect ratio (Fig. 1). Clearly, the physical length of the fibre interferes with the mobility and change of shape of the FRC in this test.

The slump test is not as effective in distinguishing differences in mixture behavior attributable to fibre length as the V-B and I.C. tests.

4 Workability Measurements for Glass and Polypropylene FRC

In the case of glass the reinforcing elements are multifilament strands which are not intended to separate into individual fibres during mixing. In the case of polypropylene they are fibrillated tapes which are intended to separate into more or less individual fibres during mixing. The complex geometry of both these types of reinforcing elements coupled with the great increase in aspect ratio that accompanies any separation makes it impossible to assign values of aspect ratio to them. Fibre length is the only readily determinable fibre characteristic, and the limited data in Fig. 4 show that fibre length is again a major factor in determining relationships between fibre content and slump, V-B time or I.C. time. Both the V-B test and the I.C. test are effective in distinguishing the effect of fibre length. However, for the 57 and 64 mm polypropylene lengths, many I.C. tests at the higher fibre contents were judged invalid because of wrapping of the fibre around the internal vibrator and/or failure of the cone to empty completely due to the high degree of cohesion of the mixture. Such difficulties were not encountered with the V-B test. Again, slump was not very effective in distinguishing the influence of fibre length on workability.

5 Effect of Concrete Mixture Parameters

Fundamentally, workability exclusive of the effect of fibre parameters is a function of the degree particle interference and the associated degree of overfilling of the voids by cement paste coupled with the fluidity of the paste.

In the first investigation reported, the minimum void content achievable with dry fine and coarse aggregates was established as 22.3% using rodded unit weight measurements. The aggregate blend was 45% fine and 55% coarse. Cement paste of water-cement ratio 0.4 was added in volume fractions calculated as 28.6% (voids overfilled by about 6%) to 34.6% (voids overfilled by about 12%). Steel wire fibres of size 25x0.4 mm were added in increments of 20 kg/m^3 or 0.25% by volume. The relationships between workability and fibre content show an apparent dependence on paste volume fraction (Fig. 5). However, this can also be interpreted as a dependence on water content. This is expected for concretes with or without fibres, so the importance of paste volume fraction *per se* is unclear.

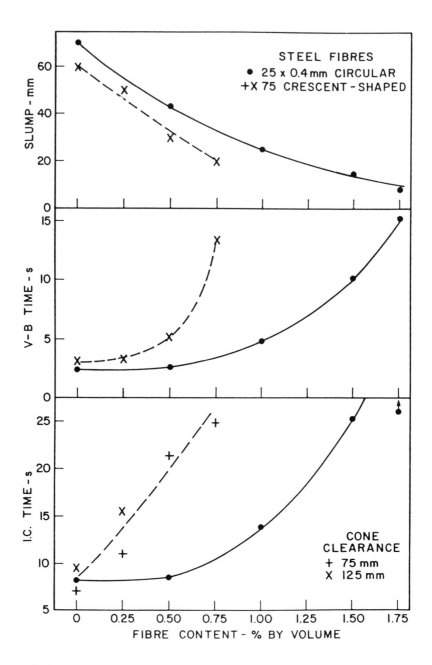

Fig.3. Effect of fibre length on workability for two types of steel fibre.

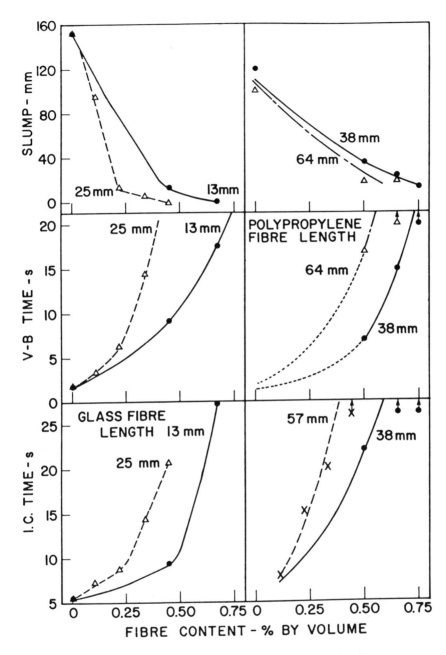

Fig.4. Effects of fibre type and length on workability for
glass and polypropylene fibres.

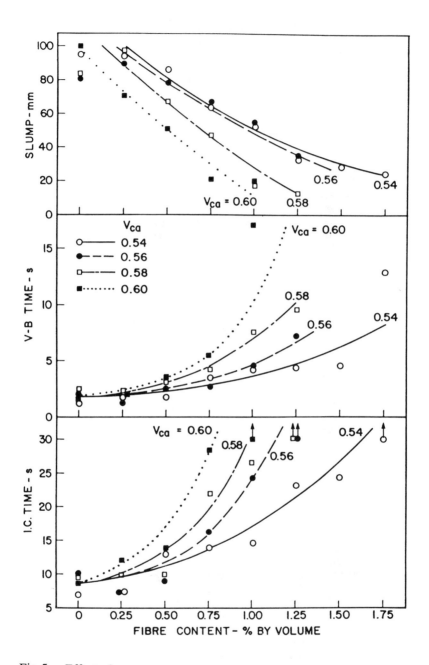

Fig.5. Effect of cement paste volume fraction and water content on workability for 25x0.4 mm steel fibres.

In the second investigation, the volume fraction of coarse aggregate (solid + voids) based on rodded unit weight was varied from the value of 0.60 which is the approximate value from the mixture proportioning tables in ACI 211.1 (ACI 1988) for the size of coarse aggregate and fineness modulus of the fine aggregate used. The water-cement ratio was 0.4. The water content was constant at 192 kg/m^3. The fibres were again 25x0.4 mm steel wire. The relationships between workability and fibre content (Fig. 6) show that decreasing the volume fraction of coarse aggregate, and therefore increasing the volume fraction of fine aggregate while keeping cement paste volume and water-cement ratio constant, improves the workability attainable especially for the higher fibre contents. Both the V-B and I.C. tests are effective in distinguishing the dependence of workability on coarse aggregate volume fraction. Clearly, a reduction in the coarse aggregate volume fraction normally obtained from ACI 211.1 tables is appropriate for a well-proportioned FRC.

6 Inverted Cone Test Variables

The clearance between the bottom of the cone and the catchment container originally specified in 1983 was 75 mm, and in the 1986 version of ASTM 995 was revised to 100 mm. In a comparison of I.C. times for 75, 100, and 125 mm clearances based on five tests at each clearance, one group observed a definite decrease in I.C. time with increase in the clearance (Table 2). They also generated some precision data showing an average coefficient of variation of 11.0% (Table 1), somewhat inferior to that reported in Table 1. Another group performing single tests at each clearance on a larger number of different mixtures confirmed the same trend (Table 3). Indeed, the averages representing the differences in I.C. time attributable to cone clearance are very similar and show that I.C. times for a 100 mm or 125 mm clearance are 81 to 83% or 73 to 74% respectively of the values for a 75 mm clearance.

Both groups considered the effect of blocking the bottom of the cone with a wood insert during filling with FRC to prevent the possibility of loss of material from the test sample prior to the start of vibration. Their results were inconclusive in showing whether this might significantly alter the measured I.C. time. Differences attributable to blocking were generally ± 3s for I.C. times of approximately 15s.

7 Conclusions

Both the V-B test and the inverted cone test are much more effective than the slump test for distinguishing the influence of fibre and matrix parameters on the workability of FRC and its placeability under vibration. The useable time range for the inverted cone test, approximately 5 to 30s, is larger than for the V-B test, approximately 2 to 15s, so the time can be determined more precisely especially for short times. However, determination of the test end-point in the inverted cone test becomes difficult or impossible with long fibres in general because flow from the cone is never completed, and with long flexible fibres like fibrillated polypropylene in particular which tend to wrap around the internal vibrator. Determination of the end-point in the V-B test does not seem to pose any special difficulty for the relatively wide range of fibre and mixture parameters examined.

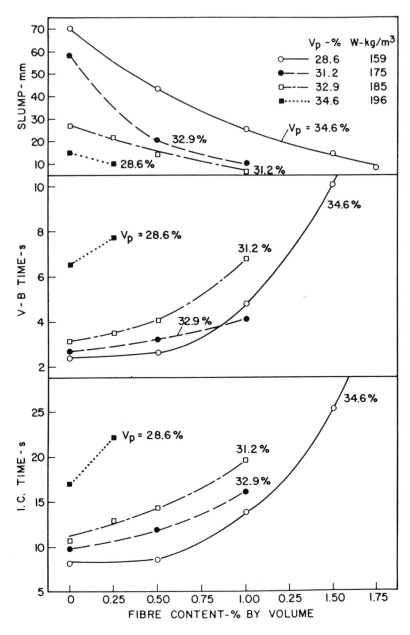

Fig.6. Effect of coarse aggregate volume fraction on workability for 25x0.4 mm steel fibres

Table 2. Inverted cone time table data for various test conditions for 0.5% of 25x0.4 mm steel fibres

	Clearance Between Cone and Bucket				
	75 mm	100 mm	125 mm	100 mm	100 mm[a]
	18.7	14.1	16.3	14.8	15.4
	21.0	15.2	18.1	15.4	13.0
	23.0	18.0	13.4	16.5	12.8
	18.5	15.5	13.0	17.1	12.5
	<u>17.8</u>	<u>17.6</u>	<u>11.2</u>	<u>16.7</u>	<u>13.2</u>
Mean	19.8	16.1 (81)[b]	14.4 (73)[b]	16.1	13.4
Std. Dev.	2.2	1.7	2.8	1.0	1.2
CofV-%	10.9%	10.3%	19.2%	6.0%	8.7%

[a]Cone blocked during filling
[b]Percent of value for 75 mm clearance condition

Table 3. Inverted cone time data for various mixtures and test conditions

Steel Fibres		I.C. Time for Cone Clearance			Paste
Type	kg/m³	75 mm	100 mm	125 mm	Volume %
25x04. mm	0	11.3	9.4 (83)[a]	8.6 (76)[a]	31.2
wire	20	12.9	9.5 (74)	8.8 (68)	
	40	14.4	9.5 (66)	9.0 (63)	
	80	19.5	13.5 (69)	9.7 (50)	
25x0.4mm	0	11.0	10.1 (92)	8.9 (81)	32.1
wire	40	13.0	12.7 (98)	10.5 (81)	
	80	18.3	16.2 (89)	15.5 (85)	
75 mm	0	9.8	9.0 (92)	7.2 (74)	32.9
crescent-	20	13.1	25.3 (≠)	12.6 (96)	
shaped	40	*	*	22.5	
75 mm	0	9.5	8.2 (86)	7.0 (74)	34.6
crescent-	20	15.5	11.9 (77)	11.0 (71)	
shaped	40	*	*	21.4	
	60	*	*	24.8	
		Mean	(83)	(74)	

[a]Percent of value for 75 mm clearance condition
*Invalid test because cone did not empty completely
≠ Probably an erroneous result

The relationship between inverted cone and V-B times seems relatively independent of many fibre and mixture parameters as both Fig. 1 and Fig. 7 indicate that inverted cone time is approximately 3.7 times the corresponding V-B time, at least for steel fibres.

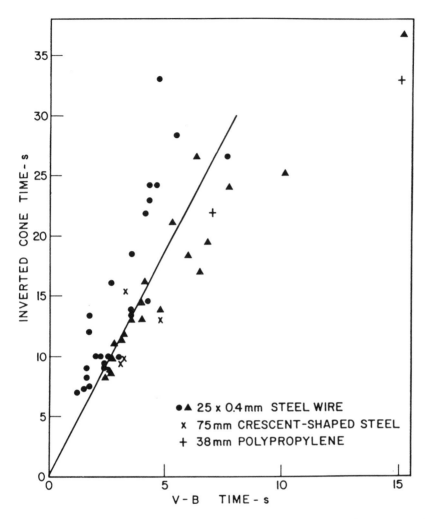

Fig.7.   Relationship between I.C. time and V-B time for a range of mixture and fibre parameters.

8 References

ACI 211 (1988) Standard Practice for selecting proportions for normal, heavyweight, and mass concrete, American Concrete Institute, Manual of Concrete Practice, Report 211.1.

ACI 544 (1978) Measurement of properties of fiber reinforced concrete, American Concrete Institute Journal, Proceedings, Vol. 75, 7, 283-289.

Johnston, C.D. (1984) Measurements of the workability of steel fiber reinforced concrete and their precision, ASTM Cement, Concrete and Aggregates, Vol. 6, 2, 74-83.

PART FIVE

HIGH STRENGTH, ULTRA-RAPID HARDENING AND VERY DRY CONCRETES

14 SPECIAL CEMENTS AND THEIR APPLICATIONS

A. D. R. BROWN and A. E. DEARLOVE
Blue Circle Technical Centre, Greenhithe, UK
D. JOHNSON
Pozament Ltd, Swadlingcote, UK
A. McLEOD
British Airways, London, UK

Abstract
This paper describes the use of a special quicksetting
and rapid hardening cement (PQ-X) developed by Pozament
Limited for repairing and replacing concrete pavements at
London Heathrow Airport.
This cement is based on a novel calcium sulfoaluminate
clinker (Rockfast) developed and manufactured by Blue
Circle Cement.
Keywords: Calcium Sulfoaluminate, Clinker Composition,
Formulations, PQ-X Concrete, Heathrow

1 Introduction

A characteristic of calcium sulfoaluminate (CSA) cements
has been their diversity, both in terms of their
composition and in their mode of use or application. The
main categories of existing types are Type K and CSA
additives. Type M and S are also closely associated
cement types. Reviews are given by Kalousek (1973) and
Hoff and Mather (1980).

This paper describes another patented variant of type
K clinker. In addition to describing the clinker
composition, a range of cements of which it is a
constituent are discussed.

Initial experience with Type K, Rockfast and PQ-X
cements suggests that sufficient performance, durability
and reliability exists to enable their use in
applications which will benefit from one or more of the
following properties:- (1) High early and late strength.
(2) Reduced drying shrinkage. (3) Expansion/self
stressing. (4) Non aggressive to GFRC.

2 Clinker Development

The production of this novel CSA clinker known as
Rockfast, was initially prompted by the need to develop
an improved cement for the production of coal mining

Special Concretes: Workability and Mixing. Edited by Peter J. M. Bartos. © RILEM.
Published by E & FN Spon, 2–6 Boundary Row, London SE1 8HN, 0 419 18870 3.

roadway supports. For this purpose, cement slurries are pumped underground to the point of placement. Further details are given by Long et al. (1987) and Brooks and Sharp (1990). Some existing cements were too alkaline from a safety consideration and required an 'aggregate', usually coal waste, as part of the mix. Such systems were also found to be abrasive to the slurry pumps. Replacements were therefore sought to overcome these shortcomings and an aggregate free system based on the formation of ettringite ($C_3A, 3CaSO_4, 32H_2O$) as the sole hydrate was adopted as the most suitable approach.

A sulfoaluminate clinker of high stability, consistent composition and predictable properties was essential and the other requirement was that the cement slurry could be pumped for a considerable time without setting in the supply pipelines.

Stability was required to avoid variations in performance due to changes on storage and the clinker was therefore designed to have very low free lime levels and to be devoid of highly reactive compounds such as $C_{12}A_7$. This also minimised the need for expensive retardation of the cement slurry. The composition targeted was free of reactive calcium silicates (C_3S and C_2S) in order both to reduce the alkalinity and allow the use of high purity aluminous raw materials for product consistency.

The composition finally adopted was one of several investigated and could be manufactured in a conventional rotary kiln with only slight modifications to the control procedures.

3 Rockfast Type K Composition

The closest equivalent production clinkers previously made were Type K clinkers of variable composition which are significantly higher in calcium silicates and free lime and lower in Klein's compound ($C_4A_3\bar{S}$).

Data for a low silicate Japanese product (Denka CSA) and a conventional calcium aluminate cement are presented in Table 1 for comparison. Since the Rockfast clinker formulation is deficient in both CaO and SO_3 for ettringite formation, wet slurries of the neat cement can be pumped for up to 24 hours with little detectable hydration. Having the source of CaO and SO_3 external to the sulfoaluminate clinker allows a degree of flexibility for expansive properties in non mining applications.

Reaction is initiated by splash mixing the cement slurry with an activator slurry containing calcium sulfate and hydrated lime as its reactive agents. Mixing is done in flexible plastic 'forms' which subsequently become the tunnel roofing supports following setting.

This low strength dual slurry system, designated as the Hydropack system, and a foam derivate marketed as

Hydrofoam, have also been used for cavity filling operations in mining situations.

Table 1. Chemical and Compound Composition of Rockfast Calcium Sulfoaluminate and of Other Clinkers for Comparison

	Rockfast	Type K (1)	Type K (2)	Denka CSA	Calcium Aluminate Cement
SiO_2	3.6	10.7	23.8	1.6	4.5
Al_2O_3	47.4	20.7	5.8	14.1	39.0
Fe_2O_3	1.4	1.7	1.1	0.3	16.0
CaO	38.0	46.9	61.7	51.8	38.0
MgO	0.3	1.1	2.0	1.2	0.6
SO_3	7.5	18.5	3.5	29.6	0.2
TiO_2	2.2	0.3	0.1	0.2	2.5
K_2O	0.16	0.29	0.25	0.06	0.15
LOI	0.3	0.7	0.7	0.6	–
Free Lime	0.3	0.3	8.0	18.1	0.1

Estimated Compound Composition					
C_4A_3S	57	38	10	28	0
CA	17	0	0	0	43
$C_{12}A_7$	0	0	0	0	9
CS	0.5	23	4	44	0
C_2AS	17	0	0	0	10
C_4AF	4	5	3	1	29
CT	4	0.5	0.2	0.3	4
C_2S	0	30	61	0	5
C_3S	0	1	10	6	0
CaO	0.3	0.3	8	18	0.1

4 Properties of Hydropack Roadway Support System

The properties of the mining product for which the clinker was initially developed are briefly presented in Table 2.

Table 2. Properties of Hydropack Roadway Support System

Method of Use	2 component slurries - cement + 'activator' Splash mixed in-situ in 1:1 ratio
Usable life of slurries	24 hours or up to 7 days using a retarder
Hydration Products	Ettringite ($C_3A, 3CaSO_4, 32H_2O$) as the only hydrated compound formed
Water Solids Ratio	12:1 by Volume
Aggregate used	None
Modulus of Elasticity (N/mm^2) cured at 20°C	8 at 2 hours, 420 at 24 hours, 2260 at 7 days
Modulus of Elasticity (N/mm^2) cured at 35°C	2120 at 2 hours, 900 at 24 hours, 730 at 7 days
Compressive Strengths (N/mm^2) cured at 20°C	0.3 at 1 hour, 2.5 at 4 hours, 4.0 at 24 hours

5 Novel Cements

In addition to the above use of this sulfoaluminate clinker, the properties of rapid setting and strength development and controlled dimensional properties have been utilised in several more conventional cement applications, using much lower water cement ratios.

Cements are produced by grinding the clinker to specific surface areas in excess of 400 m^2/kg **without** introducing further sulfate compounds such as gypsum as in conventional Portland cement production. The range of cement types can then be produced by blending the ground Rockfast clinker with ordinary Portland cements, anhydrite and at times pozzolanic additions or fibre reinforcing. Formulations are numerous and are dictated by the target application. This flexibility of approach is further extended by the use of integral admixtures, where appropriate, to provide improvements in such aspects as water reduction, self levelling, or set control.

To date no difficulties have been experienced when using normal commercially available admixtures with Rockfast derived blends.

High early strengths and the desired dimensional behaviour are achieved **consistently** as a result of the tightly controlled composition of the Rockfast clinker, and its relatively inert state prior to the commencement of normal hydration. High early strengths and expansion or shrinkage compensation are achieved by the formation of ettringite ($C_3A, 3CaSO_4, 32H_2O$), the deposition of which is supplemented both concurrently and subsequently by normal hydration products of the ordinary Portland cement component. This is reviewed by Mehta (1973) and Okushima et al. (1968). Indeed, it has been found that the sulfoaluminate and OPC act as mutual accelerators.

In contrast with normal OPC hydration, excess CaO in solution is mostly combined in the formation of ettringite leading to the virtual absence of free $Ca(OH)_2$ as a hydration product. This has been confirmed by XRD and SEM studies and has additional benefits in that the cements are more durable (in particular more sulfate resisting than OPC) and are less aggressive to glass fibres.

Examples of some of the products derived from Rockfast and their uses are shown in Table 3.

Table 3. **Examples of the special uses and properties of Rockfast Products**

Product	Use	Working Time Minutes	Compressive Strength (N/mm^2)			
			10-mins	30-mins	1-hr	24-hrs
Calcrete (1:1 sand)	Glass fibre reinforced concrete	5-30	-	-	-	43
Norcrete	Pipe weld concrete	5	5	-	18	37
Quickrock	Rapid repair mortar	15	-	10	24	65
Sprayrock	Fibre reinforced render	2-3	-	-	8	27
Anchor Bolt Grout	Grout	65-90	-	-	12*	40

* 4 hours

6 Requirement for a Special Cement at Heathrow

Heathrow has been used as an airfield for a number of years now and during the War it also handled military aircraft. During the 1950's most of the concrete panels in use today were laid; continual usage of the runways, taxiways and other areas has taken its toll. Panels are cracking and spalling and require regular maintenance, in some cases damage is so severe that complete replacement is necessary. C.A.A. legislation dictates that if the aircraft pavement is of poor quality, then it must be closed. This action is necessary from the Health and Safety point of view in that loose concrete lumps could puncture tyres or fuel lines.

Both the British Airways and British Airports Authority specifications for pavement quality concrete call for a minimum strength of $30N/mm^2$ before the concrete can be trafficked by aircraft. In most areas of the airfield, possession times of about 7 days are common and concrete based on RHPC or OPC can achieve the necessary strength within this time period. However, in some areas, the possession time is only 6 or 7 hours.

For example there are a limited number of aircraft parking areas at Heathrow and these cost £500 per hour during the peak periods.

There was a need, therefore, for a rapid strength development concrete which would allow complete replacement of the panels and be serviceable within the possession time. Hot asphalt rolling was considered a temporary repair whilst other repair hardening cements were either difficult to handle in practice or produced concrete with poor durability.

Eighteen months ago, the British Airports Authority drew up the specification for repairing a busy intersection at Heathrow, known as Coopers Crossing. The police advised that either a diversion for the perimeter road be built or the contract be completed over 2 weekends when full possession of the area was permissible. This latter option was attractive in terms of cost savings (no diversion and reduced security fees of £42,000).

7 Development and Trial Application of PQ-X Cement at Heathrow

To meet this type of need a special cement, marketed as PQ-X Cement, was developed by Pozament using the ground Rockfast clinker manufactured by Blue Circle Cement. The objective was to produce a cement which could be used in conventional concrete mixes but which would set rapidly and gain strength after placing.

Trials started in September 1991 at Heathrow. Five
trials later, the cement was used in some contracts at
Heathrow; for example in June 1992 during possession
times between 11.00 pm and 7.00 am, 225m^3 of PQ-X
concrete was successfully used to replace badly damaged
concrete panels. The chief problem which had to be
addressed on first using the material was the concern
shown by the contractors on handling a rapid setting
concrete.

At the time of the first trial, the working time of 60
minutes at 20°C was applicable when the cement powder
first came into contact with water or damp aggregates.
This 60 minute period, therefore, had to cover batching
of concrete, transport to site, pouring and finishing;
transport delays (e.g. rush hour or accidents) and
security delays (for access to airfield) could be lengthy
and increase the likelihood of premature setting of
concrete in a truck.

The transport problem was overcome by adding cement to
a truck which had been charged with aggregates at the
ready mix plant. The truck was therefore used as the
mixer, although it was not specifically designed for this
purpose. Adequate mixing of 4 cubic metres of concrete
was achieved in a 6 cubic metre truck. The concrete has
also been successfully batched through a ready mix plant
and with an auger mixer. The security problem was
overcome by increased communication amongst the relevant
parties. The system for using the concrete has evolved
over the various trials and contracts and it was possible
at Coopers Crossing to batch the concrete and finish the
slab within 30 minutes.

During the trial period, the "rights and wrongs" of
batching and handling a rapid setting concrete were
established. Initially, the concrete was mixed to a very
wet consistency so that it was, effectively, self
levelling. The advantages to the contractor were that
the trucks could be turned around faster and the panel
laid quickly. The disadvantages were the reduction in
early strength development (extending possession times)
and the formation of a smooth surface with poor skid
resistance.

Workability of the concrete was difficult to
characterise using the techniques available to
contractors; these were the slump test and compacting
factor. The concrete (when properly mixed) appears
visually, to have a slump of about 75mm but when actually
measured shows a slump of 25mm i.e. quite stiff concrete.
The compacting factor could be related to slump but this
relationship was applicable chiefly to Portland cement
based concretes. A more appropriate method would be the
Concrete Flow Table (B.S. 1881 part 105). However, its
use on site is limited and most contractors consider it
to be too sophisticated for regular use.

Various aggregate types and sizes were used successfully (i.e. limestone and gravel 40mm down) with fibres and air entraining agents. The use of the latter was mandatory under the B.A. and B.A.A.'s specifications for pavement quality concrete. However extensive testing has shown that the freeze thaw resistance of the untreated concrete is excellent, obviating the need for this admixture.

The size of the slabs which had to be replaced varied between 3m x 3m and 6m x 6m and were between 250mm and 400mm in depth. Although the concrete was poured as layers, "cold" joints were not produced and subsequent coring of the slab showed monolithic concrete. Strength development was excellent as shown in Table 4.

Table 4. Compressive Strengths of PQ-X Concrete

Age	Compressive Strength (N/mm^2)
3 hours	20
4 hours	25
6 hours	30
24 hours	50
28 days	60

8 Application at Heathrow

The replacement of damaged panels at Coopers Crossing was achieved over 2 weekends in October 1992 in temperatures less than zero degrees centigrade. The concrete was batched at the contractor's compound which was only 3 minutes drive from the site.

The successful completion of the contract in the allowed time period was only possible through the full co-operation of the interested parties - British Airways, British Airports Authority, Pozament, Costains (and their sub-contractors), and the Security Division of Heathrow Airport Limited.

Anyone using this type of cement for the first time would need similar co-operation to realise the full potential of the concrete.

9 Conclusions

A special calcium sulfoaluminate clinker (Rockfast clinker), initially developed by Blue Circle Cement for use in coal mining roadway supports (Hydropack System), and subsequently used to produce a series of quick-setting and/or shrinkage compensating cements (Rockfast Cements) has been suitably adapted by Pozament to produce a special cement (PQ-X) which can be used in conventional concrete.

This concrete can be mixed and placed in a conventional way but sets and gains strength more rapidly than ordinary concrete. It has been used successfully at Heathrow Airport where only short access times were available.

10   References

BROOKS SA, SHARP JH. (1990) Ettringite-based cements. Proc Int. Conf. Calcium Aluminate Cements, (Ed. RJ Mangabhai)   Pub.E & FN Spon, pp 335-349

HOFF GC, MATHER K. (1980)   A look at Type K shrinkage-compensating cement Production and Specifications. Cedric Willson Symp. on Expansive Cement.   ACI Pub. SP-64, pp 158-180

KALOUSEK GL.(1973)   Development of expansion cements, Kleins Symp. on Expansive Cement Concrete, ACI Pub. SP-38, pp 1-20

LONG GR, LONGMAN PA, GARTSHORE GC. (1987) Special cements for mining applications. Proc. 9th Int. Conf. Cement Microscopy, Reno, pp 236-246

MEHTA PK. (1973) Mechanism of expansion associated with ettringite formation.   Cement and Concrete Research. Vol.3 No.1, pp 1-6

OKUSHIMA M, CORDA R et al. (1968) Development of expansive cement with C_4A_3S clinker.   Proc 5th Int. Symp. Chemistry of Cement, Tokyo Vol. IV, pp 419-438

15 NORWEGIAN EXPERIENCE WITH HIGH STRENGTH CONCRETE

S. HELLAND
Selmer a.s., Oslo, Norway

Abstract
High strength concrete (characteristic strength > 65 MPa) and high performance concrete (water/binder-ratio < 0.40) became a standardized material in Norway in the mid 1980s. This paper describes the use of HSC/HPC in Norway with emphasis on site experience.
Keywords: High strength concrete, high performance concrete, practical experience.

1 Introduction

During the last decade, high performance concrete/high strength concrete, HPC/HSC, has established itself as a standardized material in the Norwegian building industry. In 1986 it was incorporated in the Norwegian Standard NS 3420 (materials and execution of concrete structures). Three years later detailed rules for design with concrete of characteristic cube strength up to 105 MPa (C-105), became a part of the design code NS 3473. A year later Finland worked out an appendix to their code covering grades up to 100 MPa.

Today there are no other national codes that pass the range of 55-65 MPa.

There are three main reasons for this, in the international context, unusual utilization of concrete:

- long traditions in applying plasticizers
- domestic production of silica fume
- 20 years continuous experience in construction of offshore structures

These three pillars of the present HSC technology will be described in this paper, together with a number of brief examples of structures where this material has been applied with some keywords on how the material behaved during the production phase.

Special Concretes: Workability and Mixing. Edited by Peter J. M. Bartos. © RILEM.
Published by E & FN Spon, 2–6 Boundary Row, London SE1 8HN, 0 419 18870 3.

2   Plasticizers

Thanks to a few enthusiastic engineers in the 1950s, the use of plasticizers was introduced early in Norway. These were of the lignosulphonate type and derived from the cellulose industry. The main motivation for taking these chemical additives into use was in most cases economic so as to obtain a cheaper mix. However, they did also open up the possibility of producing concrete with a lower w/c ratio and/or improved workability.

This mix design philosophy matured, and when the first concrete structure for the North Sea was constructed in the early 1970s, these additives represented an accepted technology and became an important constituent in the HSC of those days. The bulk concrete in the Ekofisk Tank was a grade 45 MPa with 4-6 litres of plasticizers per m^3.

Today, C-75 with a slump value in the range 22-24 cm and a combination of 6-8 litres of plasticizers and superplasticizers seems to have become the normal "offshore" mix. A similar dosage of additives is presently also normal for other heavy constructions such as bridges.

As a result, 96-100% of all concrete mixed in Norway contains plasticizers. Except for Japan, this number is far higher than any other country in the world. In most of the European countries these chemicals are only applied to 30-50% of the production.

3   Silica fume

Due to our cheap hydroelectric energy, Norway has become a leading nation in the production of silicon metal and ferrosilicon alloys. This industry produces as a byproduct a fly ash of micron sized spherical particles of highly refined Si_2O.

In the late 1970s this air pollution was no longer tolerated, and bag filters were installed to collect the dust. The use of this material was covered in NS 3420 about ten years ago.

To prepare for international recognition, FIP (Fédération Internationale de la Précontrainte) published in 1988 a comprehensive "state-of-the-art" report on the use of silica fume in concrete. This report forms one of the main pillars in the present CEN work to incorporate its use into the new generation of common European standards.

Close to one third of all Norwegian concrete contains an average of 3-5% of this pozzolana. The main motivation for taking this into use was in the first phase mainly economic, to save cement while producing normal low grade concrete. It is fair to state that this was a bad misuse of the material. Today, due to code regulations, increased cost and durability considerations, this has totally shifted. The present use is to add the silica to improve the properties particularly concerning deterioration.

To reduce chloride ingress to bridges, the Ministry of Transportation since 1988 has required the use of silica fume in all our bridges.

4   Research on HSC/HPC

For 20 years, the offshore concrete industry has acted as a locomotive for the whole building industry. Due to technical complexity, this activity has required an increased level of understanding both concerning design and material properties. Due to volume, the building industry has been able to raise money to perform such activities. Today all the major contracting companies have their own groups dealing with R&D both within their companies and in joint national programmes. Keywords for these efforts have been:

- control of fresh properties
- control of curing
- High Strength Concrete, material properties and design
- High Performance Concrete with regard to durability

The annual budget for these research activities has for a number of years been in the range of 2.5-3.5 million US$. The heavy involvement by the building industry has resulted in a direct communication between the institutes and the practising engineers.

Concerning HPC/HSC we have been through three basically different phases.

The first started when silica fume became readily available in the late 1970s and lasted to the mid 1980s. This period was characterised by pilot tests, basic research on low w/c material and the construction of a number of what we today must consider as full-scale test structures. At the same time, we realized that our current building practice had severe shortcomings concerning durability, in particular concerning chloride ingress in marine and highway construction. The new knowledge combined with an urgent need for improved building methods resulted in rather dramatic revisions of the codes in 1986 and 1988. Among other requirements, we got restrictions on w/c ratio. For aggressive environment the new limits became 0.45. For bridges and highway structures the requirement was further reduced to 0.40.

The general building industry was then faced with coping with a new material. As always when introducing new technology, we observed problems in the field that hardly could have been foreseen in the laboratory or detected during the relative few numbers of previous full-scale constructions. Both clients and contractors have gained considerable practical experience through this period "the hard way".

We are for the moment in the process of evaluating the experience and in particular by investigating the in-field performance of the first generation structures built a decade ago.

During the last few years the interest in HPC/HSC has increased considerably worldwide, and pilot structures in this material have been built in a number of countries. The main Norwegian contributions in the international development of this technology today are obviously our extensive practical experience and our "semi-old" HSC structures which are at our disposal for in-field investigations.

5   Codes and regulations

HPC/HSC is in the process of being accepted as a standardized material both concerning its durability properties and in structural design. The main prestandardisation documents in this process have been the ACI state-of-the-art report from 1984, the FIP sponsored First International Symposium on HSC in Norway in 1987, the successor in California in 1990 and the CEB/FIP state-of-the-art report on HSC from 1990.

The third FIP event is in the olympic city of Lillehammer, Norway in June 1993.

In Norway, we got concrete with characteristic strengths up to 105 MPa incorporated in the code of design, NS 3473, in 1989. Finland got an extension to their code covering grades up to 100 MPa one year later. In ACI 318-89 no concrete grades are specified. Although there are no limitations to the concrete strength in general, it is not intended to be applied to HSC. CEN have made a decision in principle to include HSC up to the same level as Norway in the next version of the European codes.

In addition to the codes, the oil companies have their own specifications concerning design of structures for the North Sea. The latest of these was published by Norsk Hydro in summer 1992 and will be applied for the floating tension leg structure to be placed in the Troll field. For durability reasons, this company requires a w/c+s ratio of less than 0.36.

6   Practical experience with HSC/HPC

In Norway, the main reason so far for taking this material into use has been durability considerations. There is also a dominant trend in taking advantage of the higher design strength. C-55 to C-75 is today specified in a number of structures. So far higher grades are only occasionally used out of strength considerations.

I will sum up some of the experience we have gained with this material and then illustrate the use of HPC/HSC in Norway by some selected examples.

6.1 Aggregate
For the majority of structures where HSC/HPC has been specified, an accompanying requirement on high workability due to dense reinforcement has been the normal case. The resulting demand to the aggregate procedure is then a material with low water demand to keep down the paste volume, a continuous grading to avoid segregation at high slump values and low content of fines < 0.25 mm due to the relative high content of cement and silica.

An increasing number of procedures today tailor the grading of the aggregate by the hydraulic "wet-processing method".

6.2 Cement and additives
To achieve high strength Norcem, the national cement producer, has modified their products. Today the main cement for offshore and other high quality structures has a mean 28 day strength of 65 MPa according to CEN EN-196.

However, as the w/c ratio has dropped, we have realized that the cement becomes far more sensitive and the risk of reactions similar to some kind of a "false set" might be a problem. In particular, for applications like highway pavements where transport is normally undertaken by lorries without any equipment for agitating or remixing, this has been a problem. These mixes always contain quite a lot of plasticizers. A number of these, in particular the lignosulphonate-based ones, are increasing this tendency.

Recently all the major suppliers have formulated a new generation of additives which are supposed to counteract this problem.

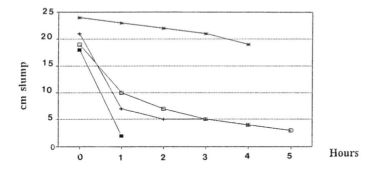

Fig. 1 Askøya Bridge. Concrete for underwater concreting. Slump loss due to 4 different plasticizers/superplasticizers w/c + s = 0.40 - 8% CSF. 10 litres of plasticizers plus superplasticizers/m^3.

The key to obtaining a low w/c ratio is a low water content. This is achieved by plasticizers. However, for all mixes there are threshold values where we hardly get any effect by increasing the amount of additives. These threshold values are certainly dependent on the aggregate, but also the cement properties. One of the main concerns in Norway is therefore to be able to understand this mechanism better and apply this knowledge into the cement production.

6.3 Mixing

All kind of mixers have been used to produce this kind of concrete in Norway. Actually the most critical factor in this process is the quality of the operators and not the mechanical equipment itself. Some types with a horizontal axis (paddle mixers), which normally are regarded as the "Rolls Royce" in the market, do not perform acceptably for concrete with slump values in excess of 22 cm as they do not introduce shear in the material. In particular for LWAC, which tend to have a very fluid initial consistency, this has created problems. One way of handling this is by adding the final portion of the water at the end of the mixing sequence.

High strength LWAC has become quite popular in bridge and offshore building during the last years. In contrast to the dominant North American tradition, the Norwegians normally use dry aggregate to improve the fire resistance and to stress the strength/density ratio. A few years ago we observed that the obvious absorption of water into the LWA in the period after mixing did cause a subsequent evacuation of air from the same particles. If this process was not followed by some kind of remixing to distribute this air, the material ended up with a rim of bubbles in the transition zone. Fig. 2 demonstrates this effect on the strength. Today remixing has become a standard procedure for this material.

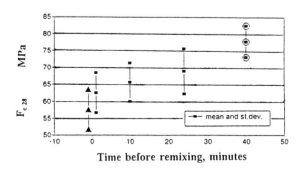

Fig. 2 Effect of remixing of LWA concrete.
Density about 1950 kg/m³.
Results from 116 full size batches.

6.4 Site equipment

The rheological properties of a highly plasticized concrete differ considerably from a normal grade mix. All the major producers of equipment are however located in countries where normal grade concrete is the only material in the market. Vibrators and slipform pavers are therefore tailored to concrete responding quite differently to an HSC/HPC. We who are daily working with these mixes in Norway are looking forward to the day one of the major producers start to optimize frequency and amplitude to our material.

6.5 Handling and curing

In addition to occasional rapid loss of workability as described above, cracking of exposed concrete has been the main problem for the contractors. Due to the absence of bleeding, the concrete tends to dry out rapidly on the surface and then menisci are formed in the top layer resulting in capillary pressure. This is the classical plastic shrinkage problem, but on these low w/c mixes the mechanism acts faster and stronger than an inexperienced crew would ever dream. To fight this problem, fog wetting or a combination of immediate sealing with a curing membrane and plastic cover has established itself as a mandatory procedure. To reduce the problem other actions introducing tension in the surface have to be controlled as well. For bridge decks, which often are inclined, the consistency of the top layer has to be stiff enough to withstand sagging. The vibrating screeds or pavers should go "uphill" and they should not do the passing late in the set period.

6.6 Competence

Working with concrete has traditionally been regarded as the opposite of "hi-tech". For this reason a too high percentage of the labour force has limited formal education and even in most universities the training of engineers for dealing with concrete on site is unsatisfactory even for normal grade concreting.

One of the main thresholds to pass to master concrete with w/c below 0.45-0.40 is certainly to educate the team involved.

The ready mix industry in Norway has faced this situation. Since the end of the 1980s every plant producing concrete for a structure classified in control class "extensive", which is the case for all HSC/HPC, needs a public certificate to document the quality of the organisation, the equipment and the formal competence of the operators and staff. These formal requirements have forced the ready mix industry back to school to learn basic concrete technology to pass the examination. This process has raised skill levels considerably, and we have even got rid of a number of rascals.

For the contractors, the client and public-induced requirements on formal competence have also gradually become stricter. To cope with this new situation, the Norwegian Association of Contractors has started a series of 1½ week training courses in concrete technology as well. We have the ambition of getting all our site teams, including those working with the vibrators, through this theoretical education within a few years. These efforts, combined with a general recession in the market which forces the unskilled labour out of this industry, have definitely improved the overall quality during the last few years.

7   Slipformed fertilizer storage silos at Herøya

Two silos for storing fertilizers were slipformed at Herøya in Norway in 1982. Each silo had a height of 28 m, diameter 27 m and wall thickness of 27 cm. The walls were heavily reinforced both with ordinary bars and post-tensioned cables. Due to the highly corrosive calcium nitrate, a dense concrete of extremely good chemical resistance had to be used. Since the paste is the weakest part of the material, the client wanted to minimize the cement and water content. At the same time the heavy reinforcement and the sensitive slipforming operation required a concrete of good workability. To meet these requirements the following mix was used:

Sulphate resistant cement	260 kg/m^3
CSF	40 kg/m^3
Total water content	112 l/m^3
Combination of melamine and lignosulphonate based plasticizers	19 l/m^3
Air entrainment agent	0.15 l/m^3
Slump	17 cm
Air content	8 %
Mean compressive strength at 28 days	65 MPa

In the mix design process we adopted the technology of Dr G. H. Tattersall to get an understanding of the rheological properties of such a mix. The results from the "two point workability tester" on three totally different mix designs, one "normal" mix (203 l water, 15 l additive, 7.5% entrained air), are shown in Fig. 3. By introducing such a high amount of plasticizers, the yield shear strength of the fresh concrete is reduced without altering the viscosity to the same degree as normally occurs when water is added to the mix to increase the slump. To get a more normal workability on the concrete, entrained air was used.

The slipforming operation was executed without any major problems in spite of the high content of CSF and plasticizers.

After 10 years of service there are no signs of deterioration in the structure, and concrete based on the above principles has since become the normal requirement in the Norwegian fertilizer industry.

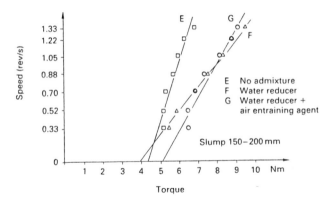

Fig. 3 Effect of combination of plasticizer and air-entraining agent

8   Underground structures in Oslo

Due to the alun shale in the central parts of Oslo, the groundwater creates a severe durability problem for concrete. Contents of up to 4 g/l SO_3 and pH values down to 2.5 have been registered. Based on in situ testing in the same area during the last 30 years, a mix based on low w/c, sulphate resistant cement and CSF has proven its superiority. Thus the bulk concrete for three of our projects in Oslo some years ago, two highway concrete tunnels and an underground parking structure for 1500 cars, were based on the following recipe:

Sulphate resistant cement	35 kg/m³
CSF	30 kg/m³
Plasticizers	8 l/m³
Entrained air	5 %
w/(c+s)	0.43
Slump	20 cm
Compressive strength at 28 days	60-75 MPa

The total quantity of concrete for the three projects was about 75 000 m³.
The main challenge on these projects was to avoid cracking in the plastic phase. Where the above mentioned precautions of immediately sealing the exposed surface by a combination of curing compound and plastic foil were taken, these problems were controlled. The other problems were cracking due to heat of hydration and subsequent cooling of massive sections restrained to neighbouring sections.

9   High strength concrete in highway pavements

In a number of the Nordic countries, almost every car is equipped with tyres having small steel studs to improve friction between the tyre and road and driver control during the winter season. These studded tyres have an enormous wearing effect on ordinary asphalt pavement. To improve the abrasion resistance, high strength concrete has been used on a number of Norwegian highways the last years. The table gives two examples of the concrete composition used on two sites during summer 1989.

Mix I is from the construction of a new length of the E-18 and E-6 highway in Norway. A total of 110 000 m^2 was placed by a slipform paver. Mix II is from a repair work on an old concrete pavement of normal quality. Six years of traffic had resulted in grooves of 35 mm depth in two of the lanes. A track with a width of 800 mm and 35 mm deep, altogether 1600 m long was milled. This track was then filled with HSC glued to the basement by epoxy.

	Mix I	Mix II
Portland cement P-30-4a kg/m^3	395	450
Silica fume kg/m^3	20	75
Plasticizer l/m^3	2	8
Superplasticizer l/m^3	6	20-25
Sand 0/8 mm, kg/m^3	913	830
Crushed stone 6/16 mm, kg/m^3		990
Crushed stone 6/22 mm, kg/m^3	1000	
w/(c+s)	0.37	0.22-0.24
Slump, cm	2-3	10
Mean strength at 28 days (cubes)	97 MPa	135 MPa

Keywords for our experience with this material during the fresh phase are:

- We have sometimes experienced a too rapid loss of workability which has caused problems when we are not able to remix the concrete on site in an "auto-mixer".
- Plastic shrinkage has however not been a big problem for pavements produced with low slump concrete.
- No slipform paver design is optimized to handle highly plasticized concrete with a continuous aggregate grading.
- The traditional latex-cement slurry often used to glue an overlay to a base concrete works as "poison" when both overlay and base concrete are low permeable HSC since the gluing effect is based on a rapid drying out of the latex.

10   Submerged concrete bridge at Karmøy, Norway

To transport gas from the oil fields of the North Sea to the exposed western coast of Norway, a double pipeline was constructed in 1982. At the shore approach the pipelines were placed inside a concrete tunnel acting as an underwater bridge over the rocky sea bed. The tunnel has a length of 590 m and was cast as five separate prefabricated elements with lengths varying from 90 to 150 m and displacement up to 7000 tons. To reduce the environmental loadings from the waves, it was essential to reduce the dimensions on the structure. This resulted in the use of high strength concrete and a very dense reinforcement (330 kg/m^3 including 80 kg/m^3 post-tensioned cables). To ensure proper concreting, a highly workable mix was needed resulting in a requirement of at least 24 cm slump. Thus the recipe for the main elements was:

OPC	400 kg/m^3
CSF	32 kg/m^3
Naphthalene based plasticizers	4 kg/m^3
Lignosulphonate based plasticizers	3 kg/m^3
w/(c+s)	0.38
Slump	24-26 cm
Mean strength at 28 days	85 MPa
Mean strength at 90 days	95 MPa

Today the Norwegian code of design allows the use of C-105 MPa concrete. In 1982 however C-65 MPa was the highest tabulated grade. For this reason the elements were designed to this quality. This requirement was met with a large margin, and the in situ strength today has obviously passed 100 MPa.

The CSF was not only used as an addition to ensure the specified compressive strength, but also to obtain the specified slump value without getting problems with bleeding and segregation.

The whole project including design and construction was completed within a period of 9 months. We happily faced no particular problems during mixing, casting or curing of the concrete on this project.

11   Gullfaks C - Offshore concrete platform in the North Sea

In May 1989 the Gullfaks C, a gravity base structure, was installed in the Norwegian sector of the North Sea. The construction of the platform, mainly by slipforming, partly took place in a dry dock, and partly floating in a fjord at the western coast of Norway. The total height of the concrete structure is 262 m, and the base area 16 000 m^2. The platform consists of 240 000 m^3 concrete with required characteristic strengths of 65 and 70 MPa. The average density of reinforcement was 290 kg/m^3.

The reason for utilizing HSC was to minimize the wall thickness and then improve the buoyancy during the tow out operation.

Including all installations, the total displacement at tow to the field was 1 500 000 tons which makes the Gullfaks C probably the largest man-made object ever to float.

The concrete for slipforming the main caissons had the following characteristics and composition:

Portland cement P-30-4A	380 kg/m^3
Superplasticizers	6 kg/m^3
Sand 0/5 mm	940 kg/m^3
w/c	0.42
Slump	24 cm
Obtained strength at 28 days on cubes:	
Mean strength	79.2 MPa
Standard deviation	3.4 MPa

12   Main girders - Stongasundet Bridge

A few years ago we constructed the Stongasundet Bridge close to the Florø in Western Norway. Due to the availability of huge floating cranes from the North Sea, we decided to construct the bridge with prefabricated main girders of up to 63 m length and weight up to 220 tons. To benefit from the high strength of the concrete and thus to increase the spans, the girders were designed with extreme slenderness. The height was up to 350 cm, the web thickness was only 26 cm. the concrete cover was 4 cm and the girders contained a number of ducts for post-tensioning. A slump value of 24-26 cm was specified to ensure a proper casting of the members. The girders were prefabricated in Bergen and transported to the site on barges. The recipe for the job was:

High strength Portland cement type Norcem P-30-4A	475 kg/m^3
CSF	40 kg/m^3
Total water content	180 l/m^3
Sand 0/8 mm	1080 kg/m^3
Gravel 8/16 mm	720 kg/m^3
Plasticizers	13 l/m^3
w/(c+s)	0.35
Obtained characteristic strength at 28 days	75 MPa

Due to the cohesive and highly fluid concrete, the slender and dense reinforced girders were cast without any honeycombing.

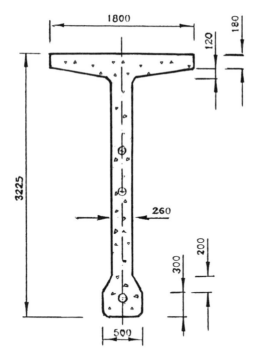

Fig. 4 Stongasundet Bridge - Main Girders
Typical cross-section - length of 63 m

13   Concrete barge "Crete Joist"

This 57 m long concrete barge was built in the UK, probably in Scotland, during
World War I, sold to Norway in the 1920s, and served along the Norwegian
coast till it was wrecked in 1942. Since then it has been exposed to tide and
waves, as well as seawater spray and frost.

The skin consists of prefabricated panels 60 cm wide and 6 cm thick.
The panels were jointed by in situ cast concrete. The reinforcement consists of
6 mm plain bars with a cover of 15-20 mm.

The construction is still in a remarkably good condition in spite of the
slender design and hostile climate. The prefabricated panels show very few
cracks and no visual evidence of corrosion resulting in cracking or spalling. The
measured electrical potential indicates, however, active corrosion, but obviously
at a very low rate.

This remarkably good performance is certainly due to the high concrete
quality and workmanship. Based on drilled out cores, the compressive strength
of the panels have been measured to 75 MPa, whereas strength levels of up to
120 MPa have been found on some of the ribs!

I would be very grateful if any of you were able to explain the technology
applied by your grandfathers to achieve such a high quality.

16 APPLICABILITY OF THE BINGHAM MODEL TO HIGH STRENGTH CONCRETE

S. SMEPLASS
Division of Structural Engineering, Norwegian Institute of
Technology, Trondheim, Norway

Abstract
This paper summarizes a minor investigation on the use of a specialized
viscosimeter on high strength concretes. The test results are compared to results
obtained for the traditional slump and flow table measures. The results indicate that
the Bingham model is highly relevant for these concrete types. Still, there is a need
for test methods to characterize other properties such as cohesion and stability.
Keywords: Workability, High strength concrete, Bingham model, Cohesivity.

Introduction

The slump measure has for some time been considered an unsatisfactory
workability criterion for high strength, high workability concretes. As a minor
subactivity within the major Norwegian research programme "High Strength
Concrete", the flow table measurement and the two point workability test were
evaluated as a supplement to the slump test /1/.

Investigations

The investigation included a total of 11 concrete mixes, all with the same cement
type, aggregate type and the same type of superplasticizer. The water/binder ratio,
w/(c+s), ranged from 0.27 to 0.50. Other variables were the paste/aggregate -
volume ratio and the content of silica fume. All concretes had a slump measure
of approximately 200 mm, obtained by varying the SP dosage according to need.

In addition to the slump measure, which was held constant as far as possible, the
workability of the concretes were characterized by the flow table measure and by
a concrete viscosimeter. The viscosimeter is a variant of Tattersalls two point
workability test, modified by Wallevik /2/. The constants g and h of the relation:

Special Concretes: Workability and Mixing. Edited by Peter J. M. Bartos. © RILEM.
Published by E & FN Spon, 2–6 Boundary Row, London SE1 8HN, 0 419 18870 3.

$$T = g + h \cdot N$$

T – torque
N – impeller speed

(1)

are considered to correspond to the yield shear stress τ_0 and the plastic viscosity μ of the Bingham model:

$$\tau = \tau_0 + \mu \cdot \dot{\gamma}$$

(2)

τ – shear stress
$\dot{\gamma}$ – the rate of shear

The constants g (Nm) and h (Nms) were determined by regression analysis of a data set consisting of 7 pairs of corresponding T and N values.

Results

Table 1 is a survey of the tested concretes and the obtained results. Fig. 1 is a plot of the corresponding g and h values for each mix. As can be seen, a reduction of the w/(c+s)-ratio in order to obtain higher strengths produces higher viscosity (h), and hence reduced workability, even if the slump is retained by the use of high dosages of superplasticizer. A conclusion is that superplasticizers primarily affects the yield shear stress, and that the effect on the viscosity is minor. This postulate has earlier been presented by Wallevik /2/.

Table 1. Test results

Mix no	Recipe	P/A	w/(c+s)	Silica fume (%)	Dosage of SP (kg/m³)	Slump (mm)	Flow (mm)	Air voids (%)	f_c (MPa)	g (Nm)	h (Nms)
41-21	ND65M	28/70	0,50	5	4,3	190	515	1,2	68,8	0,96	2,25
-22	ND95M	28/70	0,36	5	6,9	190	550	1,4	97,1	1,08	3,46
-23	ND115M	28/70	0,27	5	10,0	210	470	1,8	115,8	1,51	8,21
-24	ND95P1	30/68	0,36	5	6,7	205	555	1,5	94,4	0,88	3,08
-25	ND95P2	32/66	0,36	5	6,0	210	555	1,7	92,8	0,76	2,27
-26	ND115P1	30/68	0,27	5	9,0	220	490	1,6	110,0	1,12	5,75
-27	ND115P2	32/66	0,27	5	8,0	220	520	1,9	110,4	1,17	4,86
-28	ND95S	28/70	0,36	10	7,2	210	570	1,4	103,5	0,79	3,26
-29	ND115S	28/70	0,27	10	10,0	205	440	1,9	119,0	1,56	6,32
-30	ND95PS	30/68	0,36	10	6,0	210	620	1,4	102,0	0,69	2,30
-31	ND115PS	30/68	0,27	10	8,0	200	515	1,6	120,8	1,15	3,75

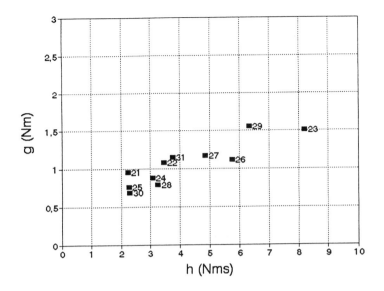

Figure 1.   Corresponding g and h values

Normally, the slump is considered to be more related to the yield shear stress (g) than the viscosity (h) of the concrete, while the flow measure is more related to the viscosity (h) than the yield shear stress (g). The present results support this theory, as illustrated in Figs 2 and 3. Since the slump is kept constant, the variation in shear yield stress (g) is relatively small. However, the viscosity (h) varies over a broad range, and so does the flow measure. Figs. 4 and 5 illustrates the effect of the cement paste / aggregate - volume ratio (P/A) on the observed yield shear stress (g) and viscosity (h). As expected, a higher content of cement paste reduces the viscosity and the need for plasticizers. According to the present results, an alternative way of reducing the viscosity of high strength concretes may be to add silica fume. As can be seen, the consumption of superplasticizer is hardly reduced either way.

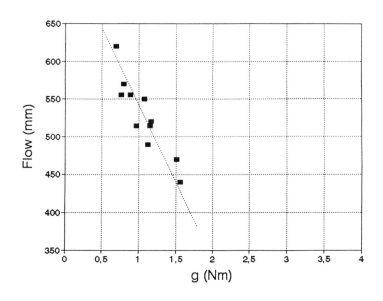

Figure 2.   Flow as a function of yield shear stress (g)

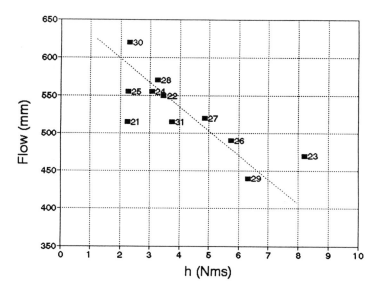

Figure 3.   Flow as a function of viscosity (h)

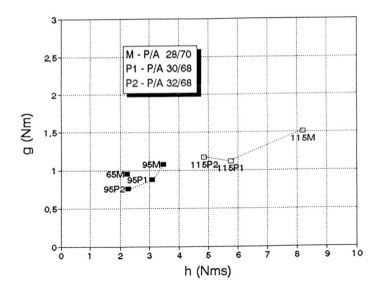

Figure 4. Effect on yield shear stress (g) and viscosity (h) of the paste-/aggregate -volume ratio

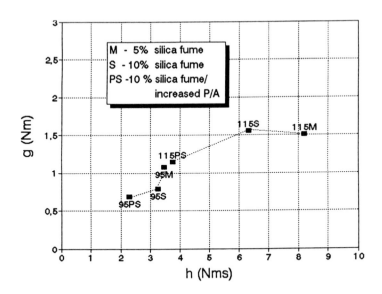

Figure 5. Effect on yield shear stress (g) and viscosity (h) of the silica fume content

Discussion

Assuming that the rheology of fresh high strength, high workability concrete is characterized by the Bingham model, and that the slump and flow measures are solely dependant on the concrete properties, the slump and flow measures might be expected to be unique functions of some combination of the g and h values. Then it would be possible to draw constant-slump-lines and constant-flow-lines in the g-h diagram, as illustrated in Fig. 6. Unfortunately, there are several indications that these relations are more complex.

As can be seen from Fig. 1, the yield shear stress (g) tends to increase with increasing viscosity, in spite of an approximately constant slump obtained through strongly increased dosages of superplasticizer. This result indicates that the slump measure not is determined by the yield shear stress and viscosity alone, but probably also by cohesion between the concrete and the slump table. The same argument goes for the flow measure. Although Fig. 3 indicates a strong correlation between the viscosity (h) and the flow value, it is obvious that a difference in flow measure of more than 100 mm for the same viscosity (h) must be explained by other factors than the minor difference in yield shear stress (g) between these two concretes.

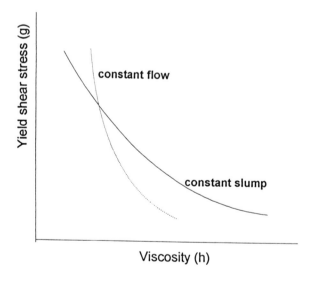

Figure 6.   *Principal correlation between the yield shear stress (g), viscosity (h), slump and flow table measurement*

Of course, this problem is not only related to empirical test methods, but also to the production properties of the concrete. For example, reduced yield shear stress and viscosity may be obtained for a high strength concrete simply by combining a superplasticizer and an air entraining agent (or by increasing the cement paste / aggregate ratio as discussed above.) However, in the high strength concrete, the air entrainment will not only reduce the viscosity, it also increases the cohessivity and stability, and reduces the response to vibration. Consequently, this air entrained high strength concrete will behave completely different from a medium strength concrete with equal yield shear stress and viscosity when placed and compacted in a heavily reinforced offshore platform shaft wall.

Conclusion

Although Binghams model may be a useful tool exploring and understanding the rheology of high strength concretes, it is clear that this model has obvious limitations. Combined use of the slump measure and the flow measure may for practical purposes provide more useful information than a more sophisticated viscosimeter. Furthermore, there is a definitive need for test methods to characterize other properties of practical importance, such as cohesion, stability and compactability (response to vibration).

References

Smeplass S: "High Strength Concrete - SP4 Materials Design - Report 4.2 Fresh Properties", Report STF65 F89001, The Cement and Concrete Research Institute, SINTEF Trondheim 1989.

Wallevik O, Danielsen SW: "Materialutvikling Høyfast Betong DP1 Tilslag - Rapport 1.1 - Statusrapport - Tilslag og fersk betongs egenskaper", STF70 A92018 SINTEF Structures and Concrete, Trondheim 1992. In Norwegian with a summary in English.

17 VERY DRY PRECASTING CONCRETE

K. JUVAS
Partek Concrete, Pargas, Finland

Abstract
This paper deals with the special properties of dry concrete, particularly its workability. The exceptional workability of dry concrete might be a disadvantage, but it can
as well create new opportunities for some products, such as
hollow-core slabs and concrete pipes. General requirements
for testing methods are also presented. Four of them are
discussed in more detail: Modified VB-test, modified Proctor test, Kango Hammer test and Intensive Compaction test.
<u>Keywords</u>: Dry Concrete, Precasting, Workability, Testing
Methods, Hollow-Core Slabs.

1 General

In some precasting processes it is normal to use a very low
amount of water in concrete. A special method is needed to
get such a concrete completely compacted. A typical water/
cement ratio for such a concrete ranges between 0.20 and
0.35 depending on the amount and quality of fines and the
admixtures used. The role of the water is mostly to take
part in the hydration reaction. Very dry concrete has also
many other names like no-slump or zero-slump concrete, low-
slump concrete, earth moist or earth stiff concrete.

Very dry concrete is being used for many purposes in the
precasting industry. Typical products are:
- extruded or slipformed hollow-core slabs
- several types of blocks, kerbs and tiles
- pipes
- slipformed filigrane slabs

In other areas of the construction industry very dry concrete has been used in roads, pavements and massive dams.
In those cases the concrete is usually called roller compacted concrete (RCC).

2 Workability

By using standardized tests or normal compaction methods
the dry concrete can be considered to have a very poor

Special Concretes: Workability and Mixing. Edited by Peter J. M. Bartos. © RILEM.
Published by E & FN Spon, 2–6 Boundary Row, London SE1 8HN, 0 419 18870 3.

workability. It cannot be compacted with a poker vibrator nor with a normal vibrating plate. It is also difficult to pump and it has a tendency to segregate. On the other hand, the cohesion of compacted concrete is very good, if the amount of fines is right. If the amount of fines is insufficient, there will be difficulties during finishing. In any case, all casting methods from transport to finishing must be correct and designed specifically for the particular workability properties of dry concrete.

3 Other properties

Why then has dry concrete been used despite the workability problems? Because the other properties are extraordinary. The cohesion and the fresh strength of newly compacted concrete is very good. Many production processes find this property necessary. It must be possible to remove pipes from their moulds right after compacting and to move them to the curing area in vertical position. Hollow core-slabs must keep their shape without any supporting mould around. It is even possible to walk after the casting machine /1/. The development of strength is very fast, and the ultimate strength is high compared with the amount of cement and other binding materials included. Of course the concrete must be properly compacted, which often is the main focus in the process and is usually related to the method of compaction. The concrete has always extra potential strength to attain if its water/cement ratio can be lowered. The matter is visualized in Fig. 1 /2/. An example of typical strength development of an extruded hollow-core slab is presented in Fig. 2. /4/
Hardened concrete with a low water/cement ratio is very dense and contains no capillary pores. This results in a good durability both against frost and chemical attack without any admixtures. Concrete is also impervious to water, and the shrinkage is slight, which will minimize the risk of cracking.

4 Testing of workability

4.1 General requirements
The term workability is very wide and can mean different kinds of properties depending on the process, product and the phase of work. The more detailed and practice based terms for different aspects of workability are
- pumpability
- cohesion
- tendency to segregation and bleeding
- compactability
- finishability

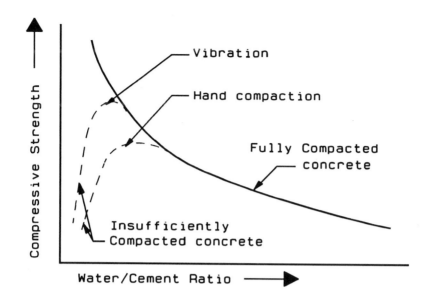

Fig. 1. The effect of water/cement ratio and different compaction methods on the compressive strength of concrete /2/

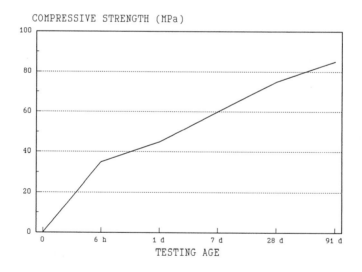

Fig. 2. The strength development of a hollow-core slab concrete. The amount of cement 360 kg/m3 and water/cement ratio 0.34 /4/.

There is probably no test method which could reliably characterize all these properties. First it must be clarified which properties are important to know in the particular case and choose the test method accordingly.

It is important that the test method imitates the real production process as far as possible. If the concrete is to be compacted by vibration, the compactability must be tested by a vibrating type of test method. For the practical user of the test results it is not necessary that these are basically rheological and theoretical. The most important is that they describe reliably the concrete and the differences between mixes in the particular process.

A test method is found suitable if:
- repeatability and reproducibility are good
- the test is fast and simple enough to be used as a process control method
- the equipment is rather inexpensive
- the method is possible to be modified according to variations in the process.

There are also different requirements that the test method and equipment must meet, such as accuracy, time demand, price and portability, depending on what the aims of the testing are:
- production quality control, on-line if possible
- development work with direct practical targets
- investigations as part of research work

The optimum compaction is an absolute condition for reaching good properties for dry concrete. It is very important to be able to test even small changes in workability. Several factors can cause changes, such as
- amount of water: added water or moisture in aggregates
- amount and fineness of fine filler
- temperature
- amount of crushed aggregate
- amount and type of cement, fly ash and other binding materials

Among conventional test methods there are very few suitable for dry concrete and even less are internationally standardized. Four test methods are presented in the following chapters:
- modified VB-test
- Proctor compaction test
- Kango Hammer test (Vibrating hammer test)
- Intensive Compaction test (IC-test)
Some further tests are mentioned in literature /3/, /7/, /13/. Due to lack of experience, they are not described here.

4.2 Modified VB-test

Origin and principle
The VB-test is developed in Sweden by V. Bährner and the
modification by CBI (Cement and Concrete Institute). It
differs from a normal VB-test by the use of two weights,
10 kg each, which make the compaction more effective and
faster. The weights press the transparent plastic plate
towards the concrete during simultaneous vibration from
below. /3/

Application
The test is applicable to dry concretes, particularly when
the compaction in the real process takes place during simultaneous vibration and pressing. This type of process is
used in the production of pipe and pavement blocks. The
original VB test is not accurate enough when the concrete
is dry, and the test takes more than 30 seconds.

Description
Normal VB-equipment with two weights of total 20 kg, hang-
ing over the plastic plate (Fig. 3 /4/) are needed. The
size of sample is about 6 litres.
The test procedure is similar to that of the normal VB-
test. The time the plate needs to get entirely in contact
with the concrete is measured.

Interpretation of the test results
The test result shows the compaction energy by vibration
and pressing that the concrete sample needs.

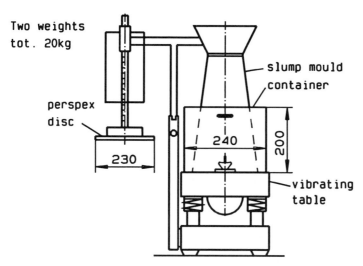

Fig. 3 Modified VB-test equipment /4/

The relationship between slump, normal VB-test and modified VB-test is shown in Fig. 4 /4/, /5/

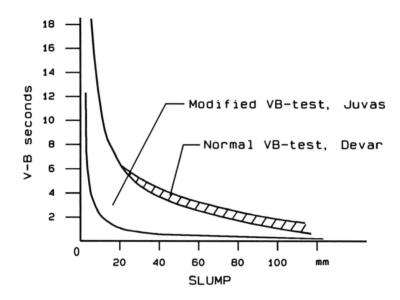

Fig. 4 Relationship between slump, normal VB-test and modified VB-test according to Devar /5 / and Juvas /4/

Precision
The timing is quite accurate, except for too flowable (less than 1 sec.) and too dry mixes (more than 15 sec.).
Very small changes in the concrete mix are not seen in the results.

Advantages
The equipment is reliable and easy to use, also in environments like concrete plants. The procedure simulates the compaction method in use for producing pipes and blocks.

Disadvantages
The test result is somewhat depending on the carefulness of the person performing the test. Very small changes in mix design will not be shown in the result.

4.3 Proctor compaction test

Origin and principle
The test has been developed in the USA and was originally used for testing different soils. The target is to determine the water content giving the best compaction with constant compacting energy by tamping.

The method is standardized at least in the USA, Germany and Sweden. /3/

Application
The test has mostly been used with lean dry concrete mixes for dams, roads, pavements and for concrete to stabilize the ground.

Description
In the modified test the sample is compacted in 5 layers, each with 25 strokes, using a 4.5 kg weight. The weight moves along a rod and the drop height is 45 cm. The principle of standard and modified test is presented in Fig. 5 /3/. The density will be calculated on the basis of the weight, volume and water content of the sample. By changing the water content it is possible to find out the maximum density and the corresponding water content. Normally four to six tests are needed. The size of sample for each test is about 3 kg.

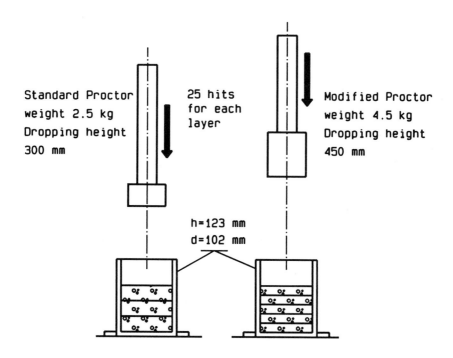

Fig. 5. Principle of the Proctor compaction test /3/

Interpretation of the test results
The test result correlates best with the properties of aggregate, such as differences in fine filler and in the shape of coarse aggregate particles. Very few comparisons between Proctor and other test methods are available. In their study Törnqvist and Laaksonen /6/ have concluded that the Proctor compaction test and the IC-test, detailed description later, show a relatively close resemblance.

Precision
The precision is good enough for field tests. There is some sensitivity depending on who performs the test; how evenly he compacts the sample by strokes.

Advantages
The test is easy and fast to perform, and the device is portable and inexpensive.

Disadvantages
This method does not consider the effect of vibration on the compaction.

4.4 Kango Hammer
Origin and principle
The method is based on the standards BS 1924:1975 "Methods of test for stabilized soils and BS 1377:1975 "Vibrating hammer method". The sample in a cubic or cylindrical steel mould is compacted by a constant pressing and vibrating force. The density of the compacted sample is then determined. The sample can be cured and compressed at later age /3/, /7/.

Application
The method has usually been used for controlling the workability of dry road concretes.

Description
The sample size is 6 to 9 kg depending on mould type. The sample is compacted in 2 to 3 layers with a vibrating hammer type Kango 900, 950 (or the like). Pressure, vibrating time and frequency are constant. The calculation of the density is similar to that used in the Proctor compaction test. The equipment is illustrated in Fig. 6. /3/,/7/.

Interpretation of the test results
The test results have been found to correlate well with the compaction test results of samples taken from road pavement works where roller compacted concrete has been used /7/.

Precision
With this test, differences can be measured to a sufficient degree for practical use.

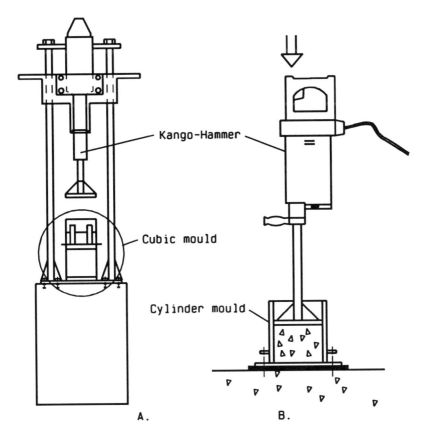

Fig. 6. Two types of Kango hammer.
        A for cubic mould /3/, B for cylinder mould /7/

5   Intensive Compaction test (IC-test)

Origin and principle
The method and equipment has been developed by I. Paakkinen
1984 in Finland. A small amount of concrete is compacted
under constant compressive force and shearing motion.

The machine measures the increasing density of the sample
during the test. The fresh strength of the cylindrical
sample can be tested immediately after stripping from the
mould, or the sample can be cured and tested later. /8/

Application
The test procedure simulates, for example, the production
of extruded hollow-core slabs. This method is applicable to
concrete of zero slump consistency or a consistency giving
Vebe times exceeding 15 to 20 seconds.

The tester is not only used for concrete quality control,
but also for developing new concrete recipes and in admixture research work. Fresh strength and strength at later
age can be measured on test specimens.
By preparing test specimens from different mixes, which the
tester always compacts in the same way, and by curing them,
it is possible to compare many properties of concrete
quickly and reliably, especially factors having effect on
structure and quality.

Description
The tester has a control unit and a main unit. The work
cylinder and a compacted sample are seen in Fig. 7.
A scale is needed as well.

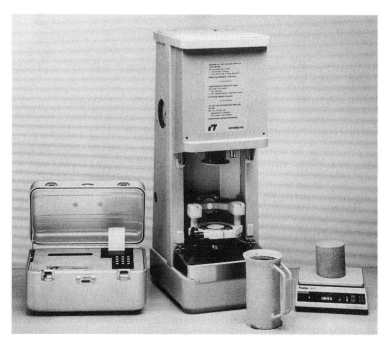

Fig. 7   Intensive Compaction Tester /8/

The different stages of the workability test are shown in
Fig. 8.

The sample is placed in the work cylinder of φ 100 mm and its weight is recorded. The normal amount of sample is 1900 to 2100 g. Test parameters are entered in the control unit. The compaction pressure and the number of compaction cycles can vary, normally they are ranging from 0.2 to 0.4 MPa (2 to 4 bar, reading in gauge) and 80 to 160 cycles, respectively.
A sample is pressed between the top and the bottom plates in the work cylinder. The top and the bottom plates remain parallel, but the angle between the plates and the work cylinder changes constantly during the circular work motion. The material moves along shear planes as it is worked. The shearing motion together with the axial compressive force allows particles to realign into more favourable positions and air is forced from the sample. The tester continually measures the height of the concrete cylinder. During testing the control unit continues to calculate density on the basis of the height and the known weight of the sample.

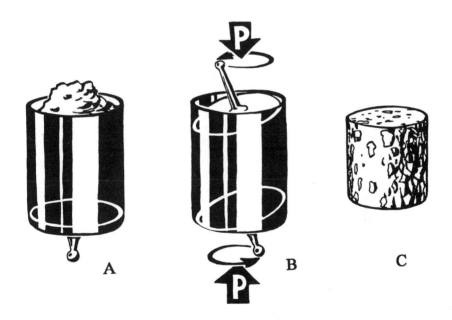

Fig. 8   Test stages /8/
A   Weighed sample in work cylinder
B   Sample under compaction
C   Sample removed from the test cylinder

After completed compaction the control unit prints out a

test report. The sample's density is printed out after 10, 20, 40, 80, 160 etc. compaction cycles. Often either 80 or 160 cycles are sufficient to draw a curve illustrating the compaction. The sample is removed from the work cylinder and can be used for further testing, e.g. fresh strength testing, or cured and tested at later age. In a research model it is possible to measure the shear resistance of concrete samples. A piece of prestressing strand is also possible to be placed in the concrete and the compacted sample used for testing the strand bond.

Another function is to use the control unit to compact the sample to a constant target density. The test result shows how many compaction cycles are required for reaching target density /9/.

Interpretation of the test results
The test results describe the energy which each concrete mix will need for complete compaction. Comparisons have been made with the Kango-hammer test /7/, Proctor test /6/, /7/ and Tattersall's 2-point test with planetary motion impeller /4/. Correlation is good with other devices except the 2-point tester. With the 2-point tester it was not possible to measure very dry concretes and to obtain reliable results.

Precision
The device is able to identify any change in the compaction properties caused e.g. by an increase or decrease of three to five litres of water per cubic meter of concrete. It does not matter whether the water originates from the aggregate or whether it has been added to the concrete mixer.

Advantages
The method is very accurate and simulates much the extruded shear-compacting process and the roller compacting process. It can be widely used for quality control, mix design and research work.

Disadvantages
The equipment is quite sophisticated and therefore rather expensive.

6 **Practical experiences with an IC-tester**

The original IC-tester and its modified version have been used in several countries from Scandinavia to New Zealand for quality control, mix design development and compaction research work. It has shown to be reliable and the results have given very usable information. Outside the precasting industry, the results have been valuable for road builders. /9/, /10/, /11/, /12/, /13/.

Examples of typical mix design tests performed at Partek
Concrete's R&D laboratory, Finland, in Fig. 9 and 10 /10/.

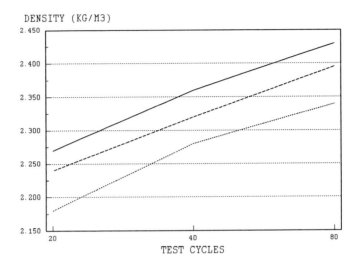

DENSITY (KG/M3)

Fig. 9   The effect of water amount on workability /10/
Legend: -   w = 115 l/m³   w/c = 0.35
         -- w = 110 l/m³   w/c = 0.33
         .. w = 105 l/m³   w/c = 0.32

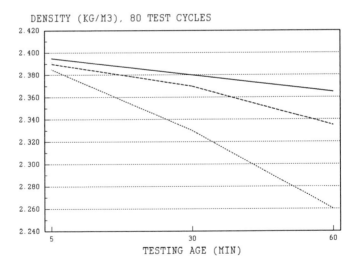

DENSITY (KG/M3), 80 TEST CYCLES

Fig. 10   The effect of concrete temperature and age on
workability /10/. Legend: Concrete temperature
-   = +20°C, -- = +30°C, .. = +50°C

7 Nordtest method for IC-tests

7.1 General

Nordtest is a joint Nordic body in the field of technical
testing. Nordtest is serving under the Nordic Council of
Ministers. The task of Nordtest is to define test methods
suitable for Nordic use - Nordtest methods. These methods
are first selected among existing standards. If no test
method is available, Nordtest develops it.

The Nordtest method for testing the compactability of concrete by using an IC-tester has been prepared and a draft
standard is being finished. The working group was formed by
specialists from laboratories in Finland, Sweden and Norway, and it consisted of those who had much experience of
practical use of the IC-tester.

The method will probably be accepted by the board of Nordtest during this year.

7.2 Results of Round Robin Test

A round robin test has recently been carried out between
ten laboratories in Finland, Sweden and Norway. The results
are now being analyzed, and will be referred to in the test
standard.

Six separate mixes were tested using two different propor-
tionings and two alternative test procedures. The first
part was carried out at constant target density (2470
kg/m^3), and the second at constant compacting energy (working pressure 4 bar, 80 compacting cycles) /14/.

An example of results from this round robin test are shown
in Table 1. The table shows that the deviations in densities are relatively small between compacted concrete samples made in different laboratories, the average deviation
ranging between 3 and 11 kg/m^3. The values for repeatability and reproducibility are also at an acceptable level.

References

1   Schwartz S. (1984) Practical hollow-core floor slab pro-
    duction below 85 dB(A), Betonwerk und Fertigteil, 12,
    807-813

2   Neville A. (1981) Properties of Concrete, London, Pitman
    779 pp

3   Andersson R. (1987) Beläggningar av vältbetong (Con-
    crete Pavements using Roller Compacted Concrete), CBI,
    Report number 87031, Stockholm, 59 pp

Table 1 IC Round Robin Test. Densities of three different mixes at ten testing laboratories. /14/

	Minimum kg/m3	Maximum kg/m3	Average kg/m3	Deviation kg/m3
Mix 4				
5 min after mixing	2.470	2.506		
10 min	2.456	2.503		
15 min	2.451	2.506		
All samples			2.484	14
Mix 5				
5 min after mixing	2.480	2.505		
10 min	2.470	2.515		
15 min	2.450	2.514		
All samples			2.486	15
Mix 6				
5 min after mixing	2.480	2.515		
10 min	2.477	2.505		
15 min	2.465	2.511		
All samples			2.487	10
Mixes 4 to 6				
All samples			2.486	12

Repeatability value r = 18

Reproducibility value R = 37

4   Juvas K. (1987) Jäykkien betonien työstettävyyskokeita (Workability tests with stiff concretes), Partek Concrete laboratory report (unpublished) Parainen, 10 pp

5   Dewar J. (1964) Relation between various workability control tests for ready mix concrete, C & CA Tech.Rep. 42, London 375 pp

6   Törnqvist J. & Laaksonen R. (1991) ICT-koe maksimikuivatilavuuspainon mittauksessa (IC-test in measuring the maximum dry weight of volume), VTT Report TGL 1880, Espoo, 17 pp

7   Magerøy H. (1987) Kontrollmetoder för valsebetong. Praktiske forsök ved Tosentunneln (Control methods for roller concrete, practical tests at the Tosentunnel), Norcem Cement A/S, Oslo, 21 pp

8   Paakkinen I. (1986) Intensive compaction tester device for testing the compactability of no-slump concrete. Nordic Concrete Research Publication No 5, Oslo, 109-116

9   Juvas K. (1988) The effect of fine aggregate and silica fume on the workability and strength development on no-slump concrete, Master's Thesis, Helsinki University of Technology, 124 pp

10   Juvas K. (1990) Experiences to measure the workability of no-slump concrete, in Proceedings of Conference of British Society of Rheology and Rheology of Fresh Cement and Concrete, University of Liverpool, London, 259-269

11   Juvas K. (1990) Experiences in measuring rheological properties of concrete having workability from high-slump to no-slump, in Proceedings of RILEM Colloquium on Properties of Fresh Concrete, University of Hannover, London, 179-186

12   Sörensen Chr. (1992) Zero-slump concrete, Thesis of Doctor, University of Trondheim, 185 pp

13   Norden G. (1992) Komprimering av torrbetong (Compaction of dry concrete) Master's Thesis, University of Trondheim, 128 pp

14   Ruohomäki J. (1993) Draft of Nordtest Method for determining the workability with IC-tester and preliminary results from round robin test, Espoo, 10 pp

18 BÉTON DE SABLE DE HAUTE RESISTANCE EN TRACTION
(High strength sand concrete in tension)

C. HUA
ENPC-CERAM – Central IV, Noisy-le-Grand, France

Abstract
The sand concrete is essentially made of sand and of cement. However, to augment his performance, diverse constituents can be incorporated (polymer, fill...). According to the idea to reinforce the contact zone between matrix (hydrates) and grains of sand, there are principally 5 possibilities. We have tried to combine these possibilities and carried out a fabrication of thin plate with sand concrete. This material elaborated has a exellent tensil strength (20 MPa).
Keywords: Concrete, Sand, Cement, Polymer, Silica fume, Clay, Contact zone, Strength.

Résumé
Les bétons de sable sont essentiellement constitués de sable et de ciment. Cependant, pour augmenter leur niveau de performance, divers composants peuvent être incorporés (polymères, fillers...). Il existe principalement 5 possibilités pour renforcer la zone de contact entre la matrice (hydrates) et les grains de sable. Nous avons tenté de combiner ces possibilités et mis au point d'un processus de fabrication de plaques minces en béton de sable. Le matériau ainsi élaboré présente une très bonne résistance en traction (20 MPa).
Mots clés: Béton, Sable, Ciment, Polymère, Fumées de silice, Argile, Zone de contact, Résistance.

1 Introduction

Les problèmes d'environnement et de raréfaction des matériaux de construction conduisent la France à s'intéresser à un matériau nouveau pour elle: Le béton de sable. Dans le cadre d'un projet national de recherche en génie civil et bâtiment SABLOCRETE, le CE-RAM a entrepris une recherche pour la mise au point de produits plats de caractéristiques mécaniques élevées en traction. L'objectif de la recherche est en fait la mise au point d'un matériau dont la résistance en traction par flexion dépasse 15 MPa et dont le coût reste à peu près 1000 francs par m^3. Un tel matériau élaboré en plaques minces de quelques millimètres d'épaisseur pourrait avoir un marché potentiel dans la préfabrication pour le bâtiment.

Les bétons de sable se distinguent des bétons classiques par leur composition. Théoriquement, il s'agit de béton constitué uniquement de sable et de ciment. Cependant, pour augmenter leur niveau de performance, divers composants peuvent être incorporés (polymères, fillers...). De plus, il sont appréciés sur le plan technique pour leurs qualités intrinsèques: bonne aptitude au pompage, propriétés autolissantes, qualités esthétiques, absence de ségrégation...

Special Concretes: Workability and Mixing. Edited by Peter J. M. Bartos. © RILEM.
Published by E & FN Spon, 2–6 Boundary Row, London SE1 8HN, 0 419 18870 3.

2 Connaissances physiques sur cette étude

2.1 La faiblesse de la zone de contact

Dans le but d'accroître les performances mécaniques des bétons, les chercheurs se sont essentiellement attachés, d'une part à améliorer les propriétés correspondantes des pâtes hydratées des ciments, d'autre part, à atteindre des compacités aussi grandes que possible.

La liaison entre la pâte de ciment hydraté et les grains enrobés n'a pas fait l'objet de recherches aussi nombreuses. Les ingénieurs comme les chercheurs ont admis, en fait, que l'adhérence était suffisante pour assurer la continuité matrice-matériaux enrobés. En effet, il est bien connu que la zone de contact (ou la zone interfaciale) est la zone la plus faible par rapport à la résistance de l'ensemble du matériau. Il est donc considéré que l'amélioration essentielle de la performance mécanique sera liée au changement de la structure interfaciale (Su, Z. et al. /1991/), sinon tous les efforts faits pour améliorer les résistances de la pâte de ciment hydraté peuvent se trouver ainsi moins efficaces que dans l'hypothèse d'une adhérence suffisante (Maso, J.C. /1982/).

En s'inspirant de cette idée sur les bétons classiques, on considère que la faiblesse de la zone de contact existe aussi entre les hydrates et les grains de sable pour le béton de sable. Ce point de vue est, en fait, confirmé par notre photo prise sur un faciès de rupture d'une éprouvette de béton de sable. On peut voir nettement apparaitre les grains de sable, qui semblent présenter une surface lisse.

2.2 Solutions pour fortifier la zone de contact

En tenant compte de la formation de la zone de contact, on peut penser à 2 facteurs qui affaiblissent le matériau:

- formation d'une couche d'eau autour des grains de sable.
- cristallisation de chaux $Ca(OH)_2$ sur les faces inertes des grains de sables.

Partant de cette idée, il existe les possibilités suivantes pour augmenter la résistance de zone de contact (Su, Z. et al. /1991/, Ohama, Y. /1987/, Zimbelman, R. /1987/):

1. Ajouter du polymère
   En fait, pour le béton sans polymère, à cause de l'effet de paroi, une zone de contact riche en eau se forme dès le malaxage. Quand on ajoute du polymère, vu que les particules polymères sont très fines, elles sont capables de remplir les espaces interstitiels. Cela peut diminuer l'épaisseur de la zone de contact grâce au meilleur empilement des particules et à la haute surface spécifique du polymère. Le polymère augmente surtout beaucoup l'adhérence de la matrice sur les grains.

2. Introduire une réaction chimique ou physique entre les grains et le ciment
   On peut remplacer les grains inactifs par des grains actifs ou activer la surface des grains (broyage par exemple).

3. Ajouter un agent tensio-actif
   Ce type de produits peut diminuer la tension superficielle de l'eau. Vue que la couche d'eau, autour du grain, produit une structure poreuse de la zone de contact, l'agent tensio-actif diminue l'intéraction des molécules d'eau, cela favorise la diminution de l'épaisseur de la couche d'eau. Par conséquent, la structure interfaciale devient plus dense.

4. Ajouter des produits réagissant entre les grains et les hydrates
   Il est possible de remplacer l'hydroxide de calcium $Ca(OH)_2$ qui se forme sur la face des grains de sable par le silicate de calcium qui a une affinité physique plus forte avec la face des grains de sable de type quartz. En plus de cela, on peut envisager une réaction entre le silicate de calcium et l'hydroxide de calcium (chaux) en formant des hydrates supplémentaires. Les fumées de silice peuvent jouer ce rôle.

5. Cure d'hydratation

D'après la littérature, la cure d'hydratation a une influence importante sur les propriétés mécaniques du béton, surtout quand on ajoute du polymère dans le béton. Il est évident que, pour la plupart des mortiers et bétons contenant des polymères, la résistance optimale est obtenue, tout d'abord en plaçant le béton frais dans un environnement humide pour atteindre un degré important d'hydratation du ciment, puis ensuite en favorisant la polymérisation par des conditions extérieures sèches (Ohama, Y. /1987/).

Le plan expérimental peut donc se diviser en 3 étapes:

- Etape 1: on combine les possibilités citées précédemment pour choisir une composition optimale du matériau en tenant compte du rapport résistance/coût.
- Etape 2: lorsque la composition est choisie, on joue sur la cure d'hydratation pour améliorer les propriétés mécaniques sans changer la composition du matériau.
- Etape 3: on effectue une fabrication systématique pour confirmer les résultats sur une quantité importante d'éprouvettes et voir des effets sous des conditions différentes.

3 Fabrication des plaques minces en béton de sable

3.1 Le matériel utilisé
- Balance SARTORIUS d'une précision de 0.5g
- Malaxeur à pâles HOBART
- Fouet électrique MOULINEX
- Malaxeur à rouleaux SCAMIA
- Plaques en aciers $360 \times 360 \times 8$, cadre en duralium 280×320
- Presse chauffante SCAMIA
- Presse INSTRON 6022 (pour des essais en flexion)

3.2 Les composants du béton de sable
- Ciment: ciment portland CPA 55 (coût 1200 F/T).
- Sable: sable de fontainebleau à granulométrie monodisperse ne contenant pas de fine. Ce sable est propre et non pollué (coût 120 F/T).
- Fine: 2 types sont utilisés, soit l'argile (kaolinite) (coût 1400 F/T), soit les fumées de silice (coût 3000 F/T).
- Polymère: Poval 205 S. C'est un polyvinyle d'acétate et d'alcool saponifié à 89 % (PVA). Ce polymère se présente sous la forme d'une poudre blanche inférieure à 100 mesh (coût 18000 F/T).
- Fludifiant: SIKA 10 (coût 8000 F/T).

3.3 Préparation de la pâte

Dans un premier temps, il faut déterminer les proportions de chacun des composants. Nous avons essentiellement fait varier la quantité de fine (fumées de silice) et d'eau, car nous considérons que ces deux facteurs sont très importants. Quant à la quantité de polymère ou de ciment, limitée par le coût, nous n'avons pas trop de choix.

A propos de la quantité de fumées de silice, nous l'avons faite varier de 30kg à 230kg pour une quantité de ciment fixée de 350kg. les résultats montrent qu'au dessus de 50kg de fumées de silice, la croissance de la résistance n'est pas intéressante au regard du prix. En tenant compte du coût de l'argile (kaolinite), nous avons fixé le créneau entre 50kg et 100kg puisque l'argile est deux fois moins cher que les fumées de silice.

Sur la quantité d'eau, les résultats montrent que le créneau optimal de E/(C+F) est entre 0.26 et 0.3 (E: eau, C: ciment , F: fines). Pour les faibles teneurs en eau, le mélange est trop sec, cela donne un matériau manquant de cohésion, mais pour les fortes teneurs en eau, les résistances du matériau ne sont pas satisfaisantes. Nous avons donc approximativement la composition suivante:

Ciment CPA 55:	350 kg
Sable de fontainebleau:	1300 - 1350 kg
Fines (fumées de silice):	30 - 50 kg
ou (argile:	50 - 100 kg
Polymères Poval 205 S:	20 kg
Fluidifiant Sika 10:	12.3 kg
Eau E/(C + F):	0.26 - 0.3

Deux modes de préparation ont été effectuées:

1. Le PVA est mélangé à sec avec le ciment, le sable et les fines. On ajoute ensuite l'eau et le fluidifiant.
2. Le mélange sec ciment-sable-fines est préparé de la même façon que précédemment. Le PVA est préparé en solution séparément et on l'ajoute au mélange sec ciment-sable-fines. On ajoute ensuite de l'eau (pour compléter la quantité d'eau nécessaire) et le fluidifiant.

   La préparation de la solution de PVA est relativement fastidieuse, car le PVA, au contact de l'eau, a tendance à former des grumeaux très difficiles à éliminer par la suite. La meilleure solution consiste, en fait, à verser progressivement tout en mélangeant le PVA dans l'eau. La solution de PVA ressemble à une mousse blanche qui "tombe" petit à petit et devient un liquide légèrement visqueux. Il s'ajoute facilement au mélange sec ciment-sable-fines.

La pâte ainsi obtenue est passé au malaxeur à rouleaux de façon à réaliser un malaxage à haut cisaillement. Ce malaxage est relativement facilité par la consistance de la pâte: lisse mais non gluante.

3.4 Fabrication de plaques minces

Le processus est le suivant: on répartit la pâte préparée dans un moule en duralium, de façon à avoir l'épaisseur de plaque homogène, on presse ensuite la pâte répartie pour avoir un matériau très dense.

En effet, la pâte ne contient que peu d'eau, car la surabondance d'eau nuit à la résistance du béton, sa maniabilité est donc très faible, il est alors difficile de répartir de manière homogène la pâte dans le moule. Si la pâte n'est pas bien répartie, on ne peut pas avoir une plaque homogène après le pressage, car certains endroits sont mieux pressés qu'ailleurs.

4 Résultats et explications

4.1 Résistance en traction par flexion

Pour effectuer la mesure de la résistance dans un essai en flexion 3-points, nous devons couper les plaques de taille $280 \times 320 \times 10$ en éprouvette de taille environ $150 \times 30 \times 10$. On charge l'éprouvette jusqu'à la rupture. Une fois la force de rupture P connue, on calcule la contrainte de rupture dite "résistance en traction par flextion" par la formule suivante (L: longeur, b: largeur, d: épaisseur):

$$\sigma_f = \frac{3PL}{2bd^2} \tag{1}$$

4.2 Résultats préliminaires

Différentes compositions du mélange de sable, ciment, fine, fluidifiant et eau, déterminées par une étude granulométrique et par le coût du mélange, ont été testées.

Les essais ont montré que l'adjonction du PVA en solution donne de meilleurs résultats que lorsqu'il est ajouté en poudre. Tout laisse à penser que le PVA se disperse moins bien dans le matériau lors d'une adjonction en poudre.

On voit également que les fumées de silice ne donnent pas de meilleurs résultats que l'argile. Ceci serait pour 2 raisons:

1. L'argile (kaolinite) que nous avons utilisée contient 70 % de SiO_2 environ, cela rend la réaction pouzzolanique possible. C'est en fait, le rôle que nous envisageons pour les fumées de silice afin de renforcer la zone de contact entre la matrice et les grains de sable. De ce point de vue, l'argile peut jouer le même rôle que les fumées de silice.

2. L'argile est aussi sous forme très fine. Le fait que nous ayons doublé la quantité d'argile en gardant à peu près le même prix du matériau, les fines remplissent mieux les pores, le matériau ainsi obtenu est donc nettement plus dense. Nous avons donc une résistance plus élevée.

Nous avons effectué aussi une cure d'hydratation sur 2 plaques de matériau fabriquées en utilisant du PVA en poudre et en solution, les plaques traitées ont été plongées dans l'eau 3 jours à l'âge de 7 jours. Ces essais ont montré que:

- Les plaques traitées donnent les meilleurs résultats.
- Le traitement diminue la différence entre l'ajout de PVA en solution et en poudre sur la résistance du matériau.

Il apparaît que la cure complète la dissolution du PVA et donc favorise la formation des films du polymère. Compte tenu des complications qu'induit l'adjonction en solution de PVA, il est nettement préférable d'apporter le PVA en poudre en le mélangeant à sec avec le ciment, le sable et les fines.

Nous avons ainsi pu atteindre une résistance de 17.8 MPa pour un béton de sable de 21 jours (14.2 MPa à 7 jours).

4.3 Fabrication systématique et expériences en série

Les résultats préliminaires nous permettent de choisir une meilleur composition et de prendre en compte le processus de traitement (cure d'hydratation) après la fabrication. Nous avons ensuite lancé une série d'essais systématiques. Pour s'assurer qu'on peut avoir de bons résultats, on a donc choisi l'adjonction de PVA en solution.

Cette campagne d'essais a été faite sur 4 plaques qui ont été chacune découpées en 3 parties égales qui correspondent aux 3 traitements différents (le tableau 1). Les tiers des

Tableau 1: Les différents traitements

	Semaine 1	Semaine 2	Semaine 3	Semaine 4	Semaine 5
Partie 1	Cellophane	Air	Air	Air	
Partie 2	Cellophane	Eau	Air	Air	
Partie 3	Cellophane	Eau	Air	Eau	Air

plaques ont été numérotés de 1 à 3 en respectant une permutation de sorte que tous ces 3 cas de traitement aient été représentés sur les 4 plaques pour tenir compte du problème d'homogénéité (la figure 1). Chaque partie est coupée en 6 éprouvettes de taille environ 130×30×10.

Les parties 1 et 2 ont été testées à l'âge de 4 semaines mais les parties 3 ont été testées à l'âge de 5 semaines pour que les éprouvettes aient le temps de sécher. Pour chaque partie (découpée en 6 éprouvettes), 3 éprouvettes sont testées sèches tandis que les 3 autres sont testées mouillées, c'est-à-dire après être restées au moins 3 heures dans l'eau.

Les résultats sont dans leur ensemble très homogènes. Il est très difficile de pouvoir distinguer des différences entre les différentes zones d'une même partie de plaque. On peut donc légitimement donner ces résultats sous forme moyenne des résistances des éprouvettes. Les moyennes sont données dans le tableau 2.

Tableau 2: Les résistances de différents traitements

No.	Partie 1		Partie 2		Partie 3	
	s	m	s	m	s	m
No. 1 (MPa)	12.33	6.15	14.86	7.16	12.84	7.62
No. 2 (MPa)	11.85	6.67	14.50	7.28	13.36	6.23
No. 3 (MPa)	12.90	5.48	14.68	6.89	13.98	7.00
No. 4 (MPa)	14.06	5.22	14.91	7.15	13.04	7.67
Moyenne (MPa)	12.76	5.88	14.73	6.87	13.30	7.17

On peut noter que les moyennes des valeurs par tiers de plaque sont elles-mêmes très homogènes, surtout pour les parties traitées (parties 2 et partie 3). Compte tenu du nombre de valeurs de départ, on peut donc considérer ces résultats comme fiables. Nous avons donc quelques commentaires suivantes:

1. De façon général, la résistance des plaques testées mouillées est environ moitié moindre que celle des éprouvettes testées séchées. Cette chute reste raisonnable par rapport au même type de résultats sur le béton classique.
2. La répétition de sécher-mouiller pour les parties 3 ne présentent pas une diminution de résistance. C'est-à-dire que les éprouvettes reprennent leur résistance quand elles se sèchent. Cela signifie que notre nouveau matériau peut être utilisé comme matériau de façade.
3. Les éprouvettes à l'âge de plus d'un mois montrent que la résistance augmente encore même au bout de quelques mois. Voici les résultats obtenus récemment (le tableau 3).

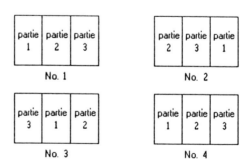

Figure 1: Découpage des plaques

174

Tableau 3: La résistance à long terme

Age	4 Semaines	4 Mois	5 Mois
Test sec (MPa)	14.73	18.74	20.23
Test mouillé (MPa)	6.87	8.01	—

Chaque chiffre est obtenu sur une dizaine d'éprouvettes. D'une part, ces résultats prouvent que notre matériau est très résistant pour un prix très intéressant: 600 F/T soit 1200 F/m^3 environ, d'autre part, le durcissement est beaucoup plus long que celui du béton classique à cause de la forte imperméabilité du matériau.

5 Conclusion

- L'objectif initial du CERAM était clair: atteindre une résistance de 15MPa avec un prix d'environ 1 000F/m^3. Nous avons mis au point un produit dont le coût est de 1 225 F/m^3 et dont la résistance est de 20 MPa. Ce résultats est très satisfaisant.
- Le processus de fabrication semble être au point. La formulation de la composition du matériau ainsi que la description du processus de fabrication nous ont paru satisfaisantes.
- Les cures d'hydratation augmentent nettement la résistance de la plaque en complétant de façon satisfaisante l'hydratation. Le traitement à l'eau renforce de plus la résistance du matériau mouillé.
- Sur le plan physique et mécanique, on considère qu'il faut renforcer la zone de contact entre la matrice et les grains de sable pour augmenter la résistance. De ce point de vue, une voie reste ouverte: ajouter un agent tensio-actif.
- Concernant la fabrication, l'adjonction de PVA en poudre reste très intéressante puisque la cure d'hydratation diminue la différence entre l'adjonction du PVA en solution et en poudre. Ceci est très important pour industrialiser le processus de la fabrication.

Remerciement

Cette recherche a été réalisée avec le soutien financier de l'association SABLOCRETE.

6 Références

Su, Z. et al. "The interface between polymer-modified cement paste and aggregates" **Cement and Concrete Research**. Vol. 21, pp. 983 - 990, 1991.

Maso, J.C. "La liaison pâte-granulats" **Le Béton Hydraulique**, Edited by Baron, J. et Sauterey, R., pp. 247 - 260. 1982.

Ohama, Y. "Principle of latex modification and Some typical Properties of latex-modified mortars and concretes" **ACI Materials Journal**, Nov.- Dec., pp. 511 - 518, 1987.

Zimbelman, R. "A method for strengthening the bond between cement stone and aggregates" **Cement and Concrete Research**, Vol. 17, pp. 651 - 660, 1987.

19 WORKABILITY AND MIXING OF HIGH PERFORMANCE MICROSILICA CONCRETE

P. L. MALE
Tarmac Topmix Ltd, Wolverhampton, UK

Abstract
A wide range of condensed fumes with varying silica contents are
produced as by-products from the high temperature reduction of quartz
in the production of silicon and its alloys. Some of these fumes,
which are called microsilica in the UK, are suitable for use in
concrete but their physical and chemical properties and the way in
which they are handled can have a considerable effect on the
performance of the hardened concrete.
Keywords: Workability, Mixing, Silica Fume, Microsilica,
Agglomeration, Workability Tests

1 Type of Microsilica and its effects on concrete

Microsilica is an incredibly fine material which must be effectively
dispersed throughout the concrete mix to make the maximum surface
area available for the hydration process. The material is extremely
difficult to handle in its as-produced state as it is so fine.
Therefore, it has to be modified in some way to enable it to be
utilised in conventional concrete batching plants. If the method of
modification involves creation of agglomeration of particles, these
then have to be broken back down in the mixing process.

A slurrified form of microsilica is available, which, if correctly
mixed at the point of production, maintains each microsphere discre-
tely in suspension, ensuring no agglomerations are created. In this
form the microsilica will disperse far easier.

If an agglomerated microsilica is incorporated into a mix, the
performance of the concrete will be affected. For example, the water
demand is increased, which, in turn, reduces the strength and the
durability. The problem of Alkali-Silica Reaction (ASR), which is
generally reduced or eliminated with microsilica can be aggravated.

The lack of understanding of the material by many researchers has
produced much data which is spurious and of little value. Most
researchers apply conventional concrete rules to their mix-designs
rather than considering the material on its own merits. For example,
dispersing admixtures work so well with the microsilica that there is
little point in not using one, and virtually all production of
microsilica concrete contains one. Fine aggregate contents also need
to be considered due to the cohesion that the well-dispersed

Special Concretes: Workability and Mixing. Edited by Peter J. M. Bartos. © RILEM.
Published by E & FN Spon, 2–6 Boundary Row, London SE1 8HN, 0 419 18870 3.

microsilica provides. Producing an ultra-cohesive mix when it is not needed is a waste of potential benefits.

2 Workability and its assessment

Workability assessment also requires special consideration. Microsilica concrete is thixotropic and consequently requires a test which involves imparting energy into the mix and then measuring how much it moves. The two-point workability method works very well and can be also used to assess other parameters such as the content of the microsilica. The slump test is inappropriate as the concrete does not slump when the cone is lifted. Consequently, for a particular measured slump the practical workability of microsilica concrete is far greater than for a conventional concrete. This is even partly true with the flow test, but for routine quality control this test is the best compromise.

Unfortunately, the slump problem is not understood by many researchers and they therefore include more water in the mix to attain a particular slump than is practically needed. Thus they report a higher water demand and reduce the quality of their hardened concrete.

Many of the problems associated with conventional concrete mixing, such as an admixture being absorbed into dry crushed rock aggregates, are aggravated in microsilica concrete because more reliance is placed on the ability of the admixture to disperse the microsilica which, in itself, provides workability. If the microsilica is not dispersed a higher water content is required to achieve the desired workability. This is another reason why higher water demands are frequently reported.

3 Mixing

It does take time for a mixer, particularly in a laboratory, to get the admixture to disperse the microsilica. Frequently, the work-ability increases before your eyes over a 5-minute interval in a laboratory mixer as the admixture gets to work.

Properly proportioned and batched, ready-mixed concrete is always stronger or requires less water than a mix produced in a laboratory. The difference is even more marked with the microsilica concrete.

I would therefore suggest that for the 'lab-crete' to truly reflect the properties of the 'real-crete' then the lab-crete should be always mixed to a lower workability.

4 Conclusions

Microsilica can be used as a quality enhancer or a cement reducer and when used as the latter, problems can be created, such as an increase in the carbonation rate and a reduction in the freeze / thaw resistance. Yet at the other end of the scale, these properties can be considerably improved.

Microsilica Concrete is frequently referred to as though it were a single product. However, depending on the relative amounts of the microsilica and cement in the mix,
the aggregate proportions,
the type and quality of the admixture,
the water content,
the quality and dispersion of the microsilica and
the method for assessment of workability,
it can be used to produce a vast range of properties.

Therefore, when considering the use of the published data on microsilica, the astute technologist should always qualify his/her comments accordingly.

PART SIX

VERY HIGHLY WORKABLE, FLOWING CONCRETE

20 DEVELOPMENT AND APPLICATION OF SUPER-WORKABLE CONCRETE

M. HAYAKAWA, Y. MATSUOKA and T. SHINDOH
Technology Research Center, Taisei Corporation, Yokohama, Japan

Abstract
Biocrete 21 was developed as an example of the super workable concrete, which has excellent deformability and high resistance to segregation, and can be filled in heavily reinforced formworks without vibrators. To study the filling ability of the super workable concrete, a new apparatus was developed. Mix proportions of Biocrete 21 could be selected through tests with this apparatus. Biocrete 21 showed excellent properties not only in fresh state but also after hardening. Some of the projects in which Biocrete 21 was employed were presented in this paper.
Keywords: Super Workable Concrete, Slump Flow, Filling Ability, Compressive Strength, Durability

1 Introduction

A super workable concrete is a new type of concrete which has excellent deformability and high resistance to segregation, and can be filled in heavily reinforced formworks without vibrators. The original idea of the super workable concrete was produced by the research team from the concrete laboratory of the University of Tokyo. The purpose of developing the super workable concrete is to construct high quality concrete structures even with unskilled labors. The super workable concrete should be excellent not only in fresh state but also in strength and durability.
"Biocrete 21" has been developed as an example of the super workable concrete in Taisei corporation utilizing a new viscosity agent made by Takeda Chemical Industries. Materials and mix proportions of Biocrete 21, evaluation of

Special Concretes: Workability and Mixing. Edited by Peter J. M. Bartos. © RILEM.
Published by E & FN Spon, 2–6 Boundary Row, London SE1 8HN, 0 419 18870 3.

workability, properties of hardened concrete, and application
to construction works are presented.

2 Biocrete 21

2.1 Materials
Biocrete 21 is produced with ordinary Portland cement, blast-
furnace slag, fly ash, aggregates, water, air entraining
water reducing agent, superplasticizer, and newly developed
viscosity agent. The viscosity agent is called biopolymer
since it is a kind of polysaccharide derived from process of
biotechnology. It is in form of white powder with specific
gravity of 1.44.

2.2 Mix Proportion
Typical mix proportion of Biocrete 21 is shown in Table 1.
Larger powder content is required than that in conventional
concrete in order to yield excellent deformability. The
combined utilization of superplasticizer and the viscosity
agent improves both deformability and resistance to
segregation of the fresh concrete.

Table 1. Mix proportions

Materials	Unit content (kg/m^3)
Water	170
Ordinary Portland cement	240
Blast-furnace slag	160
Fly ash	100
Fine aggregate	767
Coarse aggregate	850
AE water reducing agent	0.4
Superplasticizer	9.6
Biopolymer	1.0
Coarse aggregate	Maximum size=20mm
Blast-furnace slag	Blaine Fineness=4450cm^2/g
Fly ash	Blaine Fineness=3010cm^2/g
Superplasticizer	Naphtalene-base

3 Evaluation of Workability

3.1 Slump Flow

The slump flow test is conducted just same as a slump test of general concrete. In the slump flow test, consistency of concrete is expressed by the average base diameter of the concrete mass after the slump test. The slump flow value of the typical Biocrete 21 is between 60cm and 70cm (Fig.1). Since the slump flow test is easily conducted on site, it is widely used to evaluate the workability of the super workable concrete.

3.2 Apparatus for Evaluating Filling Ability

An apparatus to evaluate filling ability of concrete was developed. The apparatus is a vessel which is divided by a middle wall into two rooms shown by R1 and R2 in Fig.2. At the bottom of the wall a gate is made and installed with a sliding door. Deformed reinforcing bars with nominal diameter of 13mm are installed at the gate with center to center spacing of 50mm. This created clear spacing of 35mm between bars. To conduct the test, concrete sample is filled in R1. Small amount pressure, about 2400Pa, is applied to the concrete by putting a weight on the filled concrete in order to obtain sufficient filling height of concrete in R2 after having opened the gate. The gate is then opened by

Fig. 1. Slump flow of the Biocrete 21

185

sliding the door upward to let the concrete sample flow
through the clearance of the reinforcing steels installed at
the gate and fill the destination R2. The filling height of
the concrete in R2 is then measured and denoted as the
filling ability of the concrete.
Concretes of various mix proportions were tested with this
apparatus. Fig.3 shows the relationship between the filling
height of the concrete and slump flow value of the concrete.
When the filling height is over 30cm, it is considered that

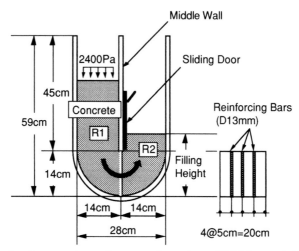

Fig.2. Apparatus for evaluating filling ability

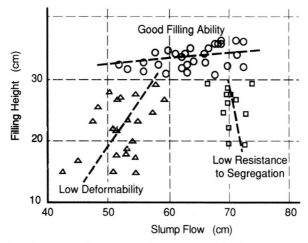

Fig.3. Relationship between filling ability and slump flow

the concrete has good filling ability. When the slump flow
is less than 50cm, the concrete will not reach the filling
height of 30cm. In this case, deformability of the concrete
is insufficient as the super workable concrete. When the
slump flow is more than 70cm, filling height of some
concretes are less than 30cm. In this case, resistance to
segregation may be insufficient. From these results, the
target range of the slump flow for the Biocrete 21 is usually
set between 60cm and 70cm.

4 Properties of Hardened Concrete

4.1 Compressive strength
Test results of compressive strength of Biocrete 21 with
different ratio among ordinary Portland cement, blast-furnace
slag, and fly ash are shown in Fig.4. Each concrete contains
500kg/m^3 of powder (ordinary Portland cement, blast-furnace
slag, and fly ash). Concrete with the higher content of
ordinary Portland cement shows the higher strength. The
compressive strength of Biocrete 21 can be controlled by
selecting proper mix proportions. Usually, the Biocrete 21
shows more than 40MPa at the age of 28days.

4.2 Durability
Since the water powder ratio of the Biocrete 21 is very low
and the water content is not so high, Biocrete 21 shows
excellent durability. Drying shrinkage of Biocrete 21 is

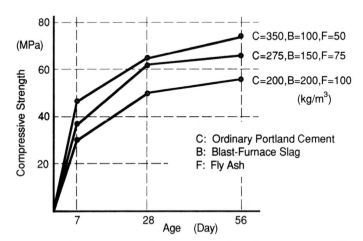

Fig.4. Compressive strength of Biocrete 21

less or equal to that of general concrete with same water content(Fig.5). Biocrete 21 with 4% or more air entrained by air entraining agents shows superior resistance to freezing and thawing(Fig.6).

5 Application

5.1 Architectural Concrete

Biocrete 21 was employed to construct a 5-story building in Tokyo(Fig.7). The outer wall of the building was artificial concrete, and its shape was rather complicated.

Biocrete 21 was produced in a ready-mixed concrete plant and transported to the site with agitating trucks. Average time for transporting from the ready-mixed concrete plant to the construction site was 1 hour, and Biocrete maintained its

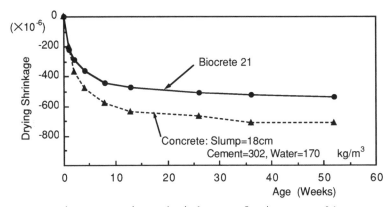

Fig.5. Drying shrinkage of Biocrete 21

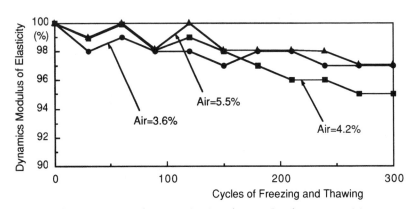

Fig.6. Freezing and thawing of Biocrete 21

large flow after transportaion. More than 3000m³ of Biocrete 21 was conveyed with a concrete pump and placed into the formworks without any vibrators.
In this project, the mix proportion was selected through tests with the apparatus for evaluating filling ability. As the quality inspection of concrete on site, slump flow tests, air content tests, and chloride content tests were conducted. Concrete with slump flow between 60cm and 75cm was placed.

5.2 Filled in steel tubular columns of a building
Biocrete 21 was also employed in the construction work of the 80-story building in Yokohama. Biocrete 21 was pumped into the steel tubular columns from basement to the 9th floor.
The typical size of the column is 600mm x 600mm, and at the each joint with beams there are diaphragm plates with a 180mm diameter hole to fill the concrete through it.
Prior to the work, Biocrete 21 was pumped into a 6m high model specimen, in which 10 diaphragm plates were set. Through this test, it was confirmed that Biocrete 21 could be filled in the tubular column without making harmful voids around diaphragm plates. Almost 900m³ of the Biocrete 21 was filled in the columns successfully.

Fig.7 Biocrete 21 was employed in outer wall of this building

6 Concluding Remarks

Biocrete 21 was developed as a super workable concrete and applied in several construction projects in Japan. The performance in fresh state is excellent due to its superior deformability and segregation resistance. Properties of the super workable concrete in fresh state can be evaluated in laboratories with the apparatus for evaluating filling ability. Test results indicated that Biocrete 21 possesses high compressive strength and excellent durability. Up to the present more than 8000m^3 of the Biocrete 21 was utilized in Japan. It is expected that the super workable concrete would make quality of concrete structures higher with less labor powers.

7 Reference

Ozawa,K.,et.al,(1989) High performance concrete based on durability of concrete, in Proceedings of the 2nd East Asia-Pacific conference on Structural Engineering and Construction, Vol.1, pp445-456

21 ASSESSMENT OF PROPERTIES OF UNDERWATER CONCRETE BY THE ORIMET TEST

P. J. M. BARTOS
Advanced Concrete Technology Group, Department of Civil
Engineering, University of Paisley, Paisley, Scotland, UK

Abstract
Workability of fresh non-dispersive underwater concretes differs very
much from that of highly workable ordinary mixes. The underwater mix
must posses a high degree of cohesion to reduce significantly the
washout when in contact with water. Simultaneously, it must have a
sufficiently high workability which guarantees a self-compaction when
placed. Common workability tests such as the Slump test or the
Spread/Flow test cannot diferentiate reliably between ordinary mixes
and nondispersive underwater concrete. The Orimet test provides easy
and reliable means of a rapid on-site identi-fication of a true
non-dispersive underwater mix. It also shows an potential to provide
an indirect measure its washout resistance.
Keywords: Underwater Concrete, Fresh Concrete, Workability, Washout
Resistance, Test Methods, Orimet, Cohesion.

1 Introduction

Placing fresh concrete under water had been undertaken by the construction industry already very early in the development of modern
concrete technology. However, the traditional mixes were highly
susceptible to the washout of the cement binder when in contact with
water. Construction techniques which permitted underwater placing
with a minimum exposure of the fresh mix to water therefore had to be
developed. A well known example is the placing by *tremie pipes* but
other systems were invented and occasionally used.

Recent advances in polymer chemistry permitted development of a
new variety of concrete admixtures specially formulated to produce a
Non-dispersive Underwater Concrete. Special underwater (U-W) admixtures increased the cohesiveness of the fresh mix so much that the
fresh mix obtained a significant degree of resistance to washout of
the cement paste when exposed to moving water.

The use of Non-dispersible Underwater concrete (NUWC) facilitates
concrete construction carried out under water or under bentonite
slurries. Easier placing of the Non-dispersible Underwater Concrete
and its washout resistance offer an opportunity to improve considerably the quality of the finished product.

The NUWC simplifies the underwater placing. In case of tremie

Special Concretes: Workability and Mixing. Edited by Peter J. M. Bartos. © RILEM.
Published by E & FN Spon, 2–6 Boundary Row, London SE1 8HN, 0 419 18870 3.

pipes it reduces the need to maintain the seal and eliminates the serious consequences of its loss. Pumping becomes easier and in appropriate conditions the NUWC can be allowed to free-fall through water without an unacceptable degree of segregation or loss of cement by the washout. However, the fresh NUWC should remain virtually self-levelling and self-compacting, both properties being highly significant in the underwater construction where access is normally very difficult and means of compaction are usually extremely limited.

Concrete construction under water is inherently difficult and all remedial works or repairs are very expensive. In such circumstances a high degree of control over the construction process and a careful maintenance of optimum properties of the fresh concrete mixes placed are essential. It is most important to *get it right first time*.

Several admixtures which impart a very high degree of cohesion and stability to a fresh mix and make it into a truly non-dispersive underwater concrete are now commercially available. A significant amount of research, namely in Japan, and a considerable practical construction experience have confirmed that NUWC mixes can produce strong and durable concrete. However, further progress in applications and a wider use of the NUWC's are hindered by a lack of standardised or at least broadly accepted methods for assessment of properties of fresh NUWC mixes on construction sites. Workability and washout resistance are particularly critical for a successful application. Such tests are also necessary for a verification of efficiency of the U-W concrete admixtures and for their rational selection.

This paper focuses on the *workability* and proposes the Orimet test as a new and more effective test method for control of workability of fresh Non-dispersible Underwater Concretes on construction sites.

2 Assessment of workability by standard tests

The current underwater concreting practice relies mainly on two methods used widely in ordinary concrete construction, namely the **Spread/Flow test** the **Slump test**. Only brief descriptions of these tests are given below. Precise data about the apparatus are indicated in appropriate national or international standards and a comprehensive survey and evaluation of the Slump test, Spread/Flow test and many other practical tests on fresh concrete has been recently provided by Bartos (1992).

2.1 The Slump test
The slump test is the most common test used for assessment of workability of fresh concrete world-wide. The range of fresh concretes for which the test provides a useful information is limited to those showing slump test results between approx. 10 mm to 150 mm. Once the workability increases to reach values greater than 150 mm the results become progressively less reliable. Further increase of workability leads to concrete slumps so great that no discernible outline of the original 'cone' remains. Then a collapsed slump, without any numerical value should be recorded. Sometimes, an overwhelming desire to obtain numerical results can lead to slump values in excess of 200mm being recorded. Such test results are meaningless as these or the

collapsed slumps cannot effectively detect differences in workability between highly workable 'flowing' mixes.

When used to assess an NUWC mix, the slump test results produce collapsed slumps or unstable slumps which increase slowly with time elapsed after the lifting of the mould. The value of the slump continues to increase for a long time, it can double its value or lead to an eventual collapsed slump.

The slump test is not suitable either for very high workability or for highly cohesive *and* highly workable underwater concretes.

2.2 Spread/Flow test
The test considered here is one of a group of 'flow' tests more appropriately described as 'spread' tests. This test fad been developed in Germany where it became standardised as a test for ordinary concrete. The test was adopted elsewhere, notably in Great Britain in the 1980s because it could be applied to flowing, super- plasticized fresh concretes which were outside the range of the slump test. The equipment required for the spread test is shown on Fig.1.

Fig 1: Spread/Flow table test equipment.

The test procedure begins by placing the mould on the engraved con-centrical ring on the top plate and filling it with concrete in a prescribed manner. The mould is then lifted and removed and the con-crete sample is allowed to slump. The top plate is then lifted to the height of the retainers without hitting them and dropped; this is repeated fifteen times.

The result of the test is found by measuring the diameter of the concrete sample, its *spread*, at the end of the test. Two perpendi-cular diameters are measured and their average is the result.

The test is susceptible to significant sources of error including the lifting action which should not cause additional jolting when the lift is restrained by the retainers, as discussed by Mor & Ravina (1986), Grube & Krell (1985). The apparatus must be set up on a firm and level sur- face, otherwise the spread caused by jolting would cause the concrete to run off the edge of the top plate. This is important in case of very highly workable concrete. Further problems are associated with the principle of the test. The jolting of the concrete test sample can initiate segregation or significant bleeding and the 'true' spread of the sample becomes difficult to measure. The test simulates poorly the actual placing of fresh concrete of high workability when any impact loading is avoided.

The basic apparatus itself is rugged, simple and portable but it is not easy to carry because of its size and weight (over 18 kg).

The Spread/Flow test is currently used for the specification of superplasticized flowing concretes which are expected to produce spreads between 50 cm to 60 cm. When applied to a fresh NUWC the results do not clearly indicate the greatly increased cohesiveness of the underwater concrete. Instead, the lower value of the spread (eg. 40cm – 50 cm) obtained is equivalent to and can be confused with that of an ordinary mix. It can show simply a result of a lower w/c or an effect of several other common factors. The actual measurement of the spread is also complicated by the continuation of the spreading, albeit at a reducing rate. It can take as much as 10 minutes before the 'final' spread is obtained.

3 The Orimet test

3.1 Principle of the test
The Orimet test was designed and developed by Bartos (1978,1982) specifically as a site test for a rapid assessment of workability of fresh flowing concretes which were outside the range of the slump test. It provides an alternative to the Spread/Flow table test.

The apparatus consists of a vertical pipe fitted with an inter- changeable orifice at its lower end. A quick-release trap door closes the orifice. The casting pipe with the trap door are supported by an integral tripod. The legs of the tripod are designed to be self- -locking when extended, securing the casting pipe in an upright position. The tripod folds back for transport and a handle fitted in a balanced position to make the whole unit easily portable.

The main parts of the Orimet are indicated in figure 2. An ordi- nary 10-litre bucket, timer or stopwatch accurate to 0.2 seconds and a cleaning brush are also required.

The design of the Orimet considered the size of the concrete sample, the length of the flow time measured and the sensitivity of the apparatus to detect changes in workability. It included the require- ments for use by a semi-skilled operator on a construction site which demanded a simple, rugged, easily portable and inexpensive equipment.

3.2 The test method
A concrete sample of at least 7.5 litres or a mass of approx. 20 kg required. Normal testing routine requires at least two, preferably three separate samples to be tested. Each sample can be retested. The

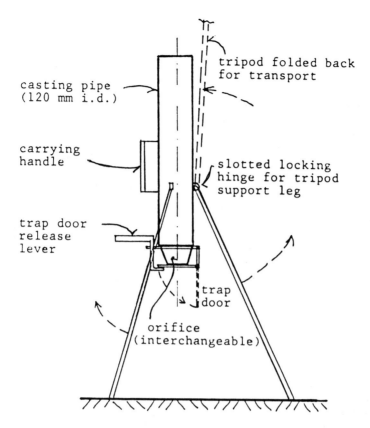

casting pipe
(120 mm i.d.)

tripod folded back
for transport

carrying
handle

slotted locking
hinge for tripod
support leg

trap door
release
lever

trap
door

orifice
(interchangeable)

Fig.2: Basic parts of the Orimet test apparatus.

test procedure consists of the following basic stages:

a. The Orimet is set up and the supporting tripod is locked in the
   extended position. The inside of the casting pipe and the orifice
   is dampened using the cleaning brush. A bucket of adequate size is
   placed under the orifice.
b. The trap door is closed and the Orimet is filled with fresh
   concrete until it is flush with the rim of the casting pipe.
   The concrete is not compacted by any means.
c. The timer is set and the timing commences simultaneously with the
   opening of the trap door which allows the concrete to flow out.
   The operator looks into the casting pipe from above and stops
   timing when light shows through.
d. The FLOW TIME (FT) is recorded to the nearest 0.2 s for times of
   up to 5 seconds and to the nearest second for the longer times.
   Mixes of insufficient workability will not flow out completely and
   light will not appear at the bottom of the casting pipe. If there
   were no visible discharge of concrete from the orifice for 60 s
   the result is recorded as a *partial flow*.

Fig.3. Orimet ready for a test.

Fig.4. Filling of the Orimet.

Fig.5. Orimet test in progress

Fig.6. Timing of the flow (FT)

The performance of the apparatus is not affected by slight deviations of the position of the casting pipe from verticality. As a guide, the max. difference of the level of concrete at the top edge of the casting pipe should not exceed 6mm.

Successive samples from one batch are tested without washing out of the apparatus if the interval between the successive tests does not exceed two minutes.

4 Workability of a Nondispersive Underwater mix

The workability tests reported below formed a part of a continuing research project on properties of NUWCs in which effectiveness of different types of the UW admixtures is being examined.

Trial mix proportions: kg/m3
      Ordinary portland cement :   450
       M graded fine aggregate :   742
   5-20mm uncrushed coarse aggregate :   900
  U-W admixture : 4.50 kg powder + 300 ml liq.
  Water / cement ratio of 0.43

The trial concrete was mixed for 45 seconds in a planetary action pan-mixer. The tests were carried out within 5 minutes after the completion of the mixing. The Orimet tester fitted with an 80mm dia. orifice was used for all tests. This size of orifice is standard for mixes with a 20mm max. size of aggregate.

4.1 Effects of dosage rate of UW admixture on the Orimet Flow Time
The dosage rate was varied from 50% to 200% of the recommended rate. The basic mix proportions remained constant.

Diagrams on Fig.7 and Fig.8 show the results. An addition of the normal dose of the U-W admixture changes the Orimet Flow Time from that corresponding to a highly workable, flowing mix to a good, cohesive non-dispersive underwater mix. Further increase in the U-W admixture dosage makes the mix exceedingly cohesive, indicated by a very long Flow Time.

The addition of the normal dose of the U-W admixture increased the Flow Time well beyond the 10 seconds which represented the upper limit of Flow Time measurable on ordinary fresh concretes. A reduction in workability of an ordinary flowing mix (FT < 3s) by a decrease of the content of a superplasticizer or by a decrease of its w/c is reflected by Flow Times rising to approx 6 to 8 seconds. **Further reduction of workability of an ordinary mix does not extend the Flow Time but produces a partial flow only.**

In case of the spread test, the normal dose of the U-W admixture brings the value of the spread down from that of a flowing mix (50-60cm range) to that of an ordinary concrete of higher workability (40-50cm range).

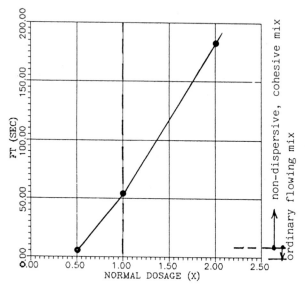

Fig.7. Effect of dosage rate of a UW admixture on the Orimet FT.

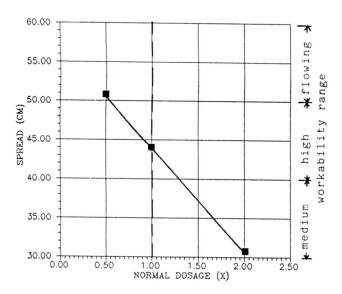

Fig.8. Effect of dosage rate of a U-W admixture on spread.

4.2 Effect of fine aggregate content.

The proportion of the fine aggregate to total aggregate content was varied from 45% (reference mix) down to 30% and up to 60%.

The results indicated the lack of sensitivity of the spread test to detect the change of the fine aggregate content which significantly affected the behaviour of the fresh NUWC mix tested. The 1/3 increase or decrease of the basic sand content produced approximate doubling or halving of the Orimet FT respectively, reflecting clearly the visibly different behaviour of the fresh mix. The same variations of the fine aggregate content did not produce significant changes of the spread test results. The spreads obtained were all within the normal scatter of results from spread tests on one mix.

Table 1. Effect of the fine aggregate content on the Orimet Flow Time FT and the spread of a fresh NUWC mix.

Admixture	UW1 (Normal Dosage)	
Results	Spread (cm)	FT (sec)
s/a = 0.30	43.00	28.80
s/a = 0.45	44.00	52.80
s/a = 0.60	40.00	117.20

5 Conclusions

The Orimet test is very sensitive to the increase of cohesion of the fresh mix caused by the addition of the 'underwater' admixture. The test results distinguish clearly between an ordinary mix and a true non-dispersive underwater mix.

The Orimet Flow Time can be used to specify FT indicating a mix with the lowest acceptable cohesiveness of a true NUWC (eg. FT > 20s). An upper limit of the FT will maintain an acceptable level of mobility and pumpability of the mix. It will also indicate a boundary beyond which the fresh NUWC will cease to be effectively self-compacting.

The specification of the maximum FT will eliminate mixes where the workability has declined below the acceptable level due to excessive time delay between mixing and placing, similar to the 'slump-loss' in ordinary mixes.

The Orimet test satisfies the criteria for a useful practical site test. It shows a good potential for development into a standard test for convenient and meaningful specification, compliance testing and site control of fresh Non-dispersive Underwater Concretes.

6 Further developments

Additional research and development project aiming to determine practical limits for workability of fresh NUWCs in terms of the Orimet Flow Times (FT) as outlined in the conclusions is being prepared.

In case of the NUWCs, the work will investigate the potentially strong relationship between the cohesiveness of the NUWCs detected by the Orimet Flow Time and the washout resistance of the mix. Preliminary tests indicated that a significant correlation existed, suggesting an additional benefit from the application of the Orimet can be obtained as a method for a rapid indirect assessment of the washout resistance of a fresh NUWC mix on a site.

7 References

Bartos, P.(1992) **Fresh concrete: Properties and Tests**, Elsevier Science Publishers, Amsterdam, 1992.

Mor, A. and Ravina, D. (1988) The DIN flow table. **Concrete International**, December 1986, 53-56.

Grube, H. and Krell, J. (1985) Prüftechnische Einflusse auf die Bestimmung des Ausbreitmasses von Beton. **Beton-technische Berichte** 84/85, Beton-Verlag, 57 - 72.

Bartos, P.(1978) Workability of Flowing Concrete and its Assessment by an Orifice Rheometer. **Concrete,** V.10, Dec.78, 28-30.

Bartos, P. Orifice Rheometer as a Test for Flowing Concrete, in **Developments in the Use of Superplasticizers** (Ed. W.M.Malhotra), American Concrete Institute SP-68, 1982, 667-682.

Acknowledgment:
The author is very grateful for the contribution to the project by Mr W Z Zhu of the China Academy of Building Materials in Beijing, China on leave at the Advanced Concrete Technology Group and Dr A.K.Tamimi of the University of Paisley.

22 DESIGN OF A RHEOMETER FOR FLUID CONCRETES

F. de LARRARD, J.-C. SZITKAR, C. HU and M. JOLY
Laboratoire Central des Ponts et Chaussées, Paris, France

Abstract
Thanks to the availability of superplasticizers, the use of fluid concretes, defined as having a slump higher than 150 mm., is more and more common. This range of concrete, while more workable than usual ones, may exhibit specific problems: rapid slump losses, unpredictable pumpability, high viscosity, proneness to segregation, difficulties in finishing slabs (especially when a slope is required) etc..
On the other hand, these concretes could be easier to characterize, because the air content is low and does not seem to depend too much on the history of the material: fluid concretes are more or less 'self-consolidating'. Therefore, one can expect to give a more rigorous description of their rheological behavior, with a suitable apparatus. Such a description would be usefull in order to simulate the fresh concrete flow (as shown by Tanigawa and col.).
The principles underlying the design of *B. T. RHEOM* will be presented, together with the current state of development. This torsonial apparatus is aimed for working in the lab and in the field, and for carrying out continuous testing of a fresh concrete specimen, with or without vibration, during the practical duration of concrete use (say, up to 2 hours).
Keywords: Bingham model, Fluid concrete, High-Performance Concrete, Rheology, Testing device.

1 Introduction

The material concrete, as a composite made up of aggregates in a matrix of silicate of hydrated lime, was known and used by the Romans. However, the secret of production of hydraulic binders was not discovered until the last century. One may therefore say that the technology of concrete dates back approximately 100 years. On the rheological plane, turn-of-the-century concretes were very dry materials that were tamped in formwork, and the worker's know-how

Special Concretes: Workability and Mixing. Edited by Peter J. M. Bartos. © RILEM.
Published by E & FN Spon, 2–6 Boundary Row, London SE1 8HN, 0 419 18870 3.

was sufficient to proportion the water to ensure good compaction. But, through a continuous process in which the cost of manpower has risen against that of the materials, there has been a trend towards materials that are closer to a concentrated suspension than to a granular packing, because the former is generally easier to place than the latter. Today, depending on the particular industrial application, materials covering the whole range of consistencies, from the "damp earth" type (for example, so-called *no-slump* concretes used for the production of concrete blocks) to the "liquid slurry" type (some building concretes, injected concretes, etc.).

Thus, given the diversity of today's concretes, it is especially important to characterize their workability properties in order to properly match the material and methods of placement.

2 Flow properties of fresh concrete and methods of measurement

With respect to the different suspensions that are the usual objects of study in rheology, the most marked characteristic of concrete is without a doubt its large grading span: coarse grains between 10 and 100 mm in diameter, small grains smaller than a micrometre (at least if the primary particles are considered).

It is the fine part of the spectrum (cement paste) that has been the object of by far the most rheological studies [Strubble 91]. Cement pastes are relatively dilute suspensions, of which the coarsest grains are close to 100 micrometres, and can therefore be tested with conventional apparatus, such as coaxial-cylinder viscometers.

Overlooking the evolving behaviour of cement pastes, related to the early chemical activity of cement in the presence of water, these materials are characterized by a roughly Binghamian behaviour, and more particularly rheothickening or rheothinning according to whether the cement concentration is high or low [Legrand 82]. The shear threshold is explained by flocculation between particles, on which it is possible to act either by adding surfactants (fluidizing admixtures or superplasticizers), or by applying vibration at relatively high frequency (typically of the order of 50 Hz, or more). Cement pastes also exhibit some thixotropy, and their structure is often modified by the phenomenon of bleeding [Legrand 82]. However, it can be stated that the behaviour of cement pastes exhibits a degree of complexity that is normal in rheology, and is therefore not too refractory to a scientific description.

The same is not true of concrete, for at least two types of reasons.

2.1 Intrinsic instability of the material when flowing
The concrete production process (mixing of constituents) leads to the incorporation of a volume of air in the mixture, which can therefore be qualified as triphasic. At the mixer outlet, a

concrete of so-called "dry" consistency can easily exhibit a swelling of the order of 10 % by volume. This porosity will be restored, in the concrete in place, to a value between 1 and 3 %. The final material - even ignoring any chemical modification - therefore differs from the initial material. Similarly, a conventional concrete of "fluid" consistency will have, at the same time, a tendency to systematic separation of its constituents, under the effect of gravity and of vibration (segregation). In this last case, at least two different materials are placed starting from a single initial mixture.

Thus, it is necessary, in order to describe the rheological behaviour of fresh concrete, to make sure that the air content does not change too much as the concrete flows and that the sample remains sufficiently homogeneous.

2.2 Crowding effects related to the limited size of the concrete part (wall effect)

It is necessary here to refer to the similarity between the fresh concrete and dry granular mixtures. The smaller ratio ϕ/d, the looser a packing of particles of diameter d in a cylinder of diameter ϕ [Caquot 35, Ben-Aïm 70]. Table 1 gives a few recent measurements illustrating this phenomenon.

Table 1: density of homometric mixtures of round aggregates compacted and vibrated in a cylinder 32 cm high (tests by LRPC of Blois, 1990). ϕ is the inside diameter of the container, d the diameter (screen mesh) of the aggregates, and c the measured density.

| | ϕ 40 | | ϕ 80 | | ϕ 160 | |
d	ϕ/d	c	ϕ/d	c	ϕ/d	c
2	20	0.6194	40	0.6155	80	0.6135
4	10	0.6126	20	0.6189	40	0.6196
8	5	0.6119	10	0.6263	20	0.6291

At the same quantity of water, a volume of concrete confined in a small container will therefore behave like a volume taken in the mass having a higher concentration of solids. Thus, to yield a representative characterization, the smallest dimension of a sample of fresh concrete must be large enough with respect to the coarsest grain of the mixture (a ratio l/d of 5 to 1 would seem to be acceptable according to the table above).

After taking into account these two major types of difficulty, inherent in the rheology of concrete, one can attempt to propose a type of modelling of the behaviour. [Tatersall 91] has shown that a Binghamian behaviour is not in contradiction with the indications of the "Two-point workability test apparatus". The test consists of measuring the relation between the torque exerted on a screw having a vertical axis rotating in a cylindrical container of fresh concrete (having the same axis) and its angular velocity. Assuming that the flow of the material is laminar, and that the shear threshold is exceeded throughout the container (no plug flow), it can be shown that the relation between the torque Γ and the velocity Ω is of the form:

$$\Gamma = g + \Omega h,$$

where g and h are constants characteristic of the material and analogous to the shear threshold and the plastic viscosity, respectively. The direct relationship with the Bingham parameters cannot however be established, because of the problems already mentioned, inherent in the rheology of concrete, together with the lack of knowledge of the stress and strain fields in the course of the test. The only advance with respect to such conventional empirical tests as the Abrams cone is to provide two quantities instead of one to qualify the rheological behaviour of the concrete. On the other hand it is impossible to give the numerical modeller a faithful constitutive law enabling him to simulate the flow of a given concrete in a structure [Tanigawa et al. 90].

More recently, [Wallevik and Gjørv 90] have proposed a coaxial -cylinder viscometer for concrete and mortar, the indications of which (torque/speed relationship) are interpreted in the same way as Tattersall's test. By calculating, analytically, the strain field in the apparatus, an attempt can be made to work back to the Bingham constant of the material [Hu et al., 93]. However, the apparatus has limitations caused by too short a distance between the cylinders (wall effect), the large probability of appearance of plug flow in the container, and bulk (impossibility of field use).

3 The needs of the engineer - Specifications of a rheometer for fluid concretes

Based on the study of the literature presented above, it is clear that there exists a range of concretes (sufficiently fluid, but not too "segregating") for which there is hope for a better description, in scientific terms, of the rheological behaviour. Such a characterization would have many applications, which might concern estimates of the following characteristics:
- the energy of mixing;
- the pumpability of the concrete, and the flowrates to be expected versus the pumping installations and the working pressures;
- the flow velocity of the material under the effect of gravity (during the emptying of mixers and transport skips, and during casting);
- the flow velocity under the combined effect of gravity and vibration;
- the time, counting from preparation, during which the

concrete may be placed;

- the maximum slope of free surfaces in fresh concrete beyond which stability problems appear (a very 'tricky' problem with fluid concretes);
- the quality of the surface of the concrete after form removal, related to the rising of air bubbles in the fresh material; etc.

3.1 Specifications

1) It must be possible to perform the test in the field, and so the sample of concrete tested must be made as small as possible, while minimizing wall effects. The minimum dimension of the sample, assuming an aggregate size of 20 mm, would thus be 10 cm.

2) It must be possible to perform shear testing of the concrete in a steady-state condition (so as to eliminate the influence of thixotropy), preferably at imposed strain rate (to avoid plug flows). Simulation of the operations of pouring and pumping involves a range of strain rates between 2 and 40 s^{-1}.

3) It must be possible to monitor the volume of the sample in the course of the test, in order to check that the occluded air content remains within a reasonable range, and that the attempt at rheological characterization is in fact meaningful. Furthermore, it cannot be ruled out that some concretes will exhibit a dilatant behaviour.

4) It must be possible to apply a mechanical vibration having a mean acceleration of 4 g rather uniformly to the sample, without interfering with the stress and strain rate measurements.

5) The test must make it possible to subject the sample to any strain history, with or without vibration.

6) Finally, for obvious practical reasons, the apparatus must be robust and not too heavy (portable by a single person).

3.2 Response to the specifications [de Larrard 90, de Larrard et al. 92]]

Constraint 2) leaves, in practice, a choice between cylinder/cylinder shear (coaxial-cylinder viscometer) or plane/plane shear. Since the first solution has shown limitations in the case of concrete, the second remains to be examined. A mechanically simple solution is to choose a movement of mutual rotation about a perpendicular axis, and to enclose the volume, two parallel planes, by a surface corresponding to the paths of the particles (i.e. two cylinders of revolution of which the axis is the same as the axis of rotation). Because of the geometry of the sample (cylinder of revolution loaded in torsion), the strain rate field is then imposed. The vertical axis entails horizontal shear planes in the fresh concrete, which serve to offset, at least partially, the effects of sedimentation by gravity. This leads to the prototype of B.T.RHEOM (see fig. 1).

This apparatus is under development. It is hoped that there will be a commercial model towards 1994-95.

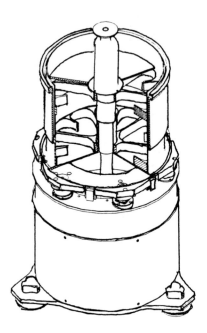

Figure 1: The prototype of B.T.RHEOM, the LCPC concrete rheometer.

4 A Method of identification of the behaviour

Let us show that with the geometry chosen, one can theoretically, from the relation $\Gamma(\Omega)$, work back to any constitutive law of the form

$$\tau = f(\dot{\epsilon})$$

(the assumption that the material follows a law of this type seems well suited to concrete at low pressure; otherwise, an internal friction component complicates the form of the constitutive law [Tanigawa et al. 87]). In the system of cylindrical co-ordinates, $\dot{\epsilon} = \dot{\epsilon}_{\theta z}$ is the strain rate and $\tau = \sigma_{\theta z}$ the shear stress.

For geometrical reasons:

$$\dot{\epsilon} = \Omega r / h.$$

The torque is then:

$$\Gamma(\Omega) = \iint \tau\, r\, ds = 2\pi \int_{R_1}^{R_2} f(\Omega r/h)\, r^2\, dr$$

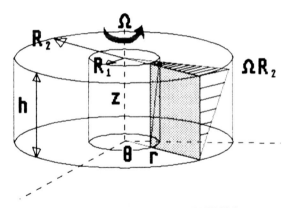

Figure 2: Theoretical rate field in B.T.RHEOM.

Setting $F(\Omega) = (3/\Omega) \Gamma(\Omega) + \Gamma'(\Omega)$ gives:

$$f(\dot{\epsilon}) = \frac{h \, \dot{\epsilon}}{2 \pi R_2{}^4} \sum_{i=0}^{\infty} k^{4i} F(k^i h \dot{\epsilon} / R_2)$$

where $k = R_1 / R_2$.

If the function Γ is sufficiently regular (and its derivative has an upper bound), the series converges quickly and it is easy, by numerical means, to obtain $f(\dot{\epsilon})$.

In the Binghamian case, the calculation is simpler. It is easily shown that the function $\Gamma(\Omega)$ is affine. If Γ_0 and Γ' represent, respectively, the ordinate at the origin and the slope of the straight line:

$$\tau_0 = \frac{3 \, \Gamma_0}{2\pi (R_2{}^3 - R_1{}^3)} \qquad \mu = \frac{2 \, h \, \Gamma'}{\pi (R_2{}^4 - R_1{}^4)}$$

where τ_0 and μ are the shear threshold and the plastic viscosity of the concrete, respectively.

5 Conclusion

A torsion viscometer is a tool theoretically well suited to study of the rheological behaviour of fluid concretes. Unlike existing apparatus, it must allow identification of a range of behaviours much broader than those following Bingham's model. For instance, concrete under vibration is unlikely to be a Bingham material. A number of difficulties may however arise at the time of development of the test (mainly the perturbation of the flow by

the friction of the vertical walls). However, even if the rheological parameters identified in the test suffer from a systematic
error, this error can be evaluated by means of finite-elements
calculations. Therefore, it is hoped that many industrial problems
that are today poorly controlled can be handled better, with
apparatus designed in response to a specification taking account
of the needs of theoretician and practitioner alike.

6 References

Ben-aïm R. (1970), "Study of the texture of packings of grains.
    Applications to the determination of the permeability of binary
    mixtures in the molecular, intermediate, laminar regime." State
    Thesis, University of Nancy (in French).
Caquot A. (1935), "Function of inert materials in concrete",
    Memorandum of the Société des Ingénieurs Civils de France,
    July-August (in French).
de Larrard F. (1990), "Thoughts concerning a new test to measure
    the consistency of concrete", Bulletin de Liaison des
    Laboratoires des Ponts et Chaussées, No. 166, March-April (in
    French).
de Larrard F., Hu C., Szitkar J.C. (1992), "Conception d'un
    rhéomètre pour bétons fluides" (Design of a rheometer for fluid
    concretes), 27° Colloque Annuel du Groupe Français de
    Rhéologie, Les Cahiers de Rhéologie, Vol. X, N° 3-4, November
    (in French).
Hu C., de Larrard F., Gjørv O.E. (1993), "Investigations on the
    viscosity of high-performance concrete", to be proposed for
    publication in Magazine of Concrete Research.
Legrand C. (1982), "The structure of suspensions of cement" and
    "The behaviour of suspensions of cement" in Le Béton
    Hydraulique, Presses de l'ENPC, Paris (in French).
Struble L.J. (1991), "The Rheology of Fresh Cement Paste".
    International Symposium on Advances in Cementitious Materials,
    American Ceramic Society, Gaitherburg.
Tanigawa Y., Mori H., Watanabe K. (1990), "Computer Simulation of
    Consistency and Rheology Tests of Fresh Concrete by
    Viscoplastic Finite Element Method", RILEM Conference
    "Properties of Fresh Concrete", Hannover, October.
Tanigawa Y., Mori H., Tsutsui K., Kurokawa Y. (1987),
    "Constitutive Law and Yield Condition of Fresh Concrete".
    Transactions of the Japan Concrete Institute, Vol. 9.
Tattersall G.H. (1991), "Workability and Quality Control of
    Concrete", E & FN Spon ed., Chapman & Hall, London.
Wallevik O.H., Gjørv O.E. (1990), "Development of a coaxial
    viscometer for fresh concrete", RILEM Conference "Properties of
    Fresh Concrete", Hannover, October.

23 SPECIFYING FLOWING CONCRETE – A CASE STUDY

D. J. CLELAND and J. R. GILFILLAN
The Queen's University of Belfast, Belfast, Northern Ireland

Abstract
A practical application of flowing concrete is described.
The need for a method of assessing the flow
characteristics is discussed and methods of achieving
this, in particular the flow table and flow trough, are
examined.
Keywords: Concrete, Flow, Workability, Testing.

1 Introduction

The ever increasing need for high levels of thermal
insulation in houses and residential accommodation has led
one UK company, at least, to look to a method of walling
which is a complete departure from the traditional masonry
cavity walls. Springvale EPS Ltd. purchased the licence
for a French system of poured concrete walls in which the
concrete is cast between permanent shutters of expanded
polystyrene. In 1988 the company joined with Queen's
University Belfast to form a Teaching Company, supported
by SERC and DTI, to adapt the system to British design and
construction
practice. The outcome, in 1992, was a certificate of
approval from the British Board of Agrement.
    Through a combination of design study, laboratory
experiments and site trials an acceptable solution was
devised. One aspect of this was the specification and
testing for compliance of the concrete.

2 KEPS walls

The arrangement for KEPS walls is shown in Fig. 1. Moulded
high density expanded polystyrene panels, 1260mm x 250mm x
50mm thick and polypropylene spacers (Fig. 2) clip
together to create a rigid formwork into which concrete is
poured. When the concrete has set the polystyrene
formwork panels remain fixed to both sides of the wall and
act as thermal insulation; giving the finished wall a high
insulation value. The 160mm concrete core of the wall is
reinforced generally to resist lateral loads and to

Special Concretes: Workability and Mixing. Edited by Peter J. M. Bartos. © RILEM.
Published by E & FN Spon, 2–6 Boundary Row, London SE1 8HN, 0 419 18870 3.

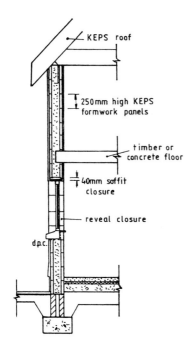

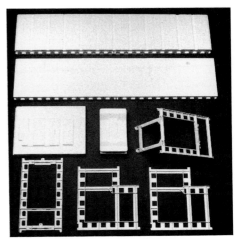

Fig.1. KEPS wall construction   Fig.2. KEPS components

control shrinkage and particularly to span windows and
door openings. The concrete mix used must be able to flow
freely around the formwork spacers and reinforcement.
There are two reasons for this:

(a)   The concrete, whether pumped or delivered by
       skip, needs to be able to disperse along the
       wall away from the point of introduction.
(b)   Vibration of the concrete to compact it is
       undesirable since damage of the formwork can
       be caused easily.

British practice differs from that in France in at
least two respects relevant to this type of construction.
Firstly cranes are not commonplace on small building
sites in Britain so pumping represents a more feasible
approach then using skips. Secondly the amount of
reinforcement required by British Codes and Building
Control is greater. Obstruction to flow is therefore more
significant.

Discussions with a local ready mix supplier led to a
mix with free flowing characteristics using gravel or
crushed aggregate of 10mm maximum size and an
aggregate/cement ratio of 3.6. A free water/cement ratio
of 0.65 is used in conjunction with a superplasticiser.

210

Trials were carried out on full scale walls. One purpose was to measure shrinkage and to assess the risk of render cracking as a result of differential movement. The second objective was to study whether or not the concrete was 'self compacting'. When all measurements were complete the formwork was removed and the concrete inspected for voids. This showed some voids in the concrete at relatively conjested locations, such as corners, when the concrete had been placed after the effect of the superplasticiser had begun to diminish. This clearly showed that although a satisfactory fluid concrete could be achieved small reductions in workability were sufficient to render the concrete unsatisfactory. This was important since it meant that:

(a)   Specifying a prescribed mix was an unwise way forward since variations in materials (e.g., aggregates) could lead to insufficiently fluid concretes in some regions of the country.

(b)   Pumpability was not an adequate test of compliance since it would be possible to pump a concrete which was not sufficiently fluid to flow freely within the formwork without entrapping voids. In any case the use of a skip to transport the concrete was not ruled out.

(c)   Achieving the correct mix proportions was not a guarantee of adequacy since an initially fluid concrete could become unsatisfactory if placing the concrete was delayed.

3 Specification

The manual for KEPS therefore adopted the approach of specifying a designed mix. Strength was relatively unimportant but a 28-day cube compressive strength of $30N/mm^2$ was chosen. On workability we were influenced by the need to specify it in terms of some parameter which was widely accepted and could be carried out easily and cheaply on site.

The first of these objectives militated in favour of the flow table test in BS1881:Part105(1984). Simple trials were carried out by observing the flow of concrete poured at different times after mixing. The flow in a section of formwork about 1m long was compared with the measured flow diameter. These results are shown in Table 1.

As a result a specification was adopted which stated that concrete, when placed should have a flow table diameter of at least 650mm. However a concern was that the flow diameter varied so little between a concrete which was more fluid than necessary and one which was not sufficiently fluid. Nevertheless still influenced by the

Table 1. Flow diameter compared to visual assessment

Flow Diameter	Slope of concrete surface in formwork	Comments
700mm	30 in 1000	Too fluid
690mm	100 in 1000	Optimum
665mm	170 in 1000	Near limit

need to have a widely accepted test this specification was included in the manual and on this basis the BBA have issued an Agrement Certificate.

In the meantime we have carried out further tests to see if some improvement can be obtained. One idea has been to use the flow table in a non-standard way; to use a smaller number of drops or none at all justified by the fact that vibration is not part of the placing process. Another avenue explored is the use of the flow trough as

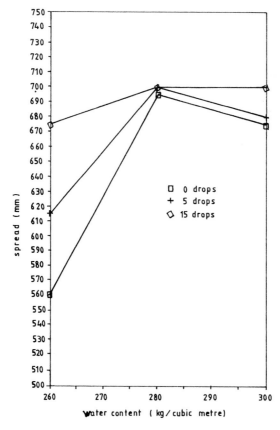

Fig. 3. spread on flow table

in DTp Standard BD27/86(1986). A special feature used in
this study was the use of perspex for the trough so that
the concrete could be examined for voids. This may not be
practical in normal situations but a transparent window
could easily be incorporated in the trough side walls.
This study is continuing but initial findings indicate
that the flow trough is a more sensitive measure of
workability. Three mixes of identical properties except
for the water content were tested. The diameter of the
spread on the flow table was measured after zero, 5, 10
and 15 drops.

The results for the flow table are shown in Fig. 3 from
which it is clear that measuring the diameter before any
drops of the table is no more or no less sensitive to
changes in workability than the diameter after 15 drops.

For the flow trough the time for the flow to reach a
450mm line and a 750mm line were measured (Fig.4). Both of
these varied quite significantly with water content and
therefore show promise as a means of specifying the flow
characteristics and testing for compliance. Further work
is ongoing to complete this study from which it is hoped
to be able to replace the flow table with a simple,
inexpensive, which can be standard equipment for all KEPS
licenced contractors.

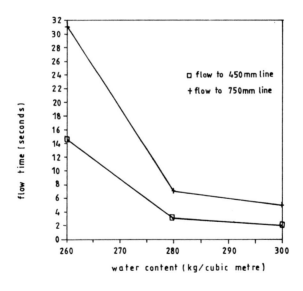

Fig. 4. flow time in flow trough

4 Conclusions

1.For a self compacting fluid concrete pumpability may not be a sufficient test of suitability.

2.The flow table is not an accurate indicator of fluidity at the level necessary for self-compacting flowing concrete.

3.The flow trough provides a method which tests the concrete under conditions which reflect the method of placing and the time for flow to travel along the trough provides a measure which appears to be sensitive to changes in water content and therefore workability.

5 References

BS1881:Part 105:1984, Testing Concrete - Method for determination of flow, British Standards Institution, 8p.
BD27/86, Materials for the repair of concrete highway structures, Department of Transport Highways and Traffic Departmental Standard, 1986, 15p.

PART SEVEN

FOAMED AND SPRAYED CONCRETE

24 FOAMED CONCRETE – MIXING AND WORKABILITY

S. KARL and J.-D. WÖRNER
Institut für Massivbau, Technische Hochschule Darmstadt,
Darmstadt, Germany

Abstract
This paper describes peculiarities of mixing and workability of two types of foamed concrete: (1) foamed concrete containing stable air cells uniformly distributed in a mix including cement paste and, if any, fine natural sand aggregate and (2) lightweight aggregate foamed concrete where the interstitial voids between the coarse lightweight aggregates are filled with a porous matrix consisting of aerated cement paste without any fine aggregates. Two principal possibilities of generating and entraining of stable air cells are discussed, and recommended mixing procedures are described. Finally, workability characteristics and possible tests for the evaluation of consistency are reported.
Keywords: Foamed concrete, Lightweight aggregate foamed concrete, Preformed foam, Mix-foaming, No-fines concrete, Foam concentrates, Yield, Batching sequence.

1 Introduction

Foamed concrete (or cellular concrete) is a lightweight cellular material produced by introducing sufficient quantities of stable air pores into net cement paste or fresh mortar consisting of cement paste and fine aggregate (the latter only for dry densities over approx. 500 kg/dm^3). In special cases, coarse lightweight aggregate can also be included, giving higher strengths as compared with foamed concrete of the same density without coarse lightweight aggregates. Normal-weight coarse aggregate is usually not used, because of the high risk of segregation due to the large differences in density of the light matrix and the aggregate grains.

This paper gives an overview over the principal possibilities of generating stable air pores within the matrix, recommended mixing procedures and characteristics of workability of two different types of foamed concrete: foamed concrete without coarse aggregate using preformed foam and mix-foamd no-fines lightweight aggregate concrete.

Special Concretes: Workability and Mixing. Edited by Peter J. M. Bartos. © RILEM.
Published by E & FN Spon, 2–6 Boundary Row, London SE1 8HN, 0 419 18870 3.

2 Production

2.1 Generation of air pores
Principally, there are two possibilities for the generation of the air pores in the concrete:

(a) adding of a stable preformed aqueous foam at the mixer and thoroughly blending into a premixed mix;
(b) mix-foaming of a mixture of water, cement, foam concentrate, and aggregates.

Method (a) is the preferred method for the production of foamed concrete in the low density range requiring large amounts of air pores. An important advantage in this field of application is that the amount of air pores can be controlled very easily, by adding a predetermined volume of preformed foam, whereas with method (b) the air content depends on many influences which cannot be kept easily under control and where the proper quantity of foam concentrate may have to be determined by trial and error for the particular mixer used and the concrete mix under consideration.

Preformed foam is made by blending foam concentrate, water and air in special foam generators. The resulting foam has a creamy consistency. One type of foam generator, preferably used at batching plants or on job sites where large amounts of foamed concrete are to be produced, is operating with compressed air and diluted foam concentrate solution (e.g. 3 % by mass). An other type, often used for feeding truck mixers on building site, is connected to the drinking water conduit, if available, or to the water tank of the truck mixer (required minimum water pressure 3 bar) and sucks in the necessary air from the atmoshere and the foam concentrate directly from the tank or barrel as delivered, due to the action of the water jet.

Very stable foams especially suitable for the production of foamed concrete are achieved by using foam concentrates on the basis of hydrolyzed proteins. Their density is normally in the range of 60 to 80 kg per m^3 provided that the foam generator is properly adjusted. Higher densities indicate that the foam might be too wet. This should be corrected by increasing the proportion of air.

The stability of preformed foam produced with concentrates on the basis of chemical detergents is not as good, in most cases, but may be also sufficient. The density of these foams may be lower, e.g. (30 to 50) kg/m^3.

Different test methods have been developed for the evaluation of the stability of the foam, e.g. a test where the time is measured required for a certain quantity of foam to flow out of a funnel with a specified outlet. Such or similar tests may be helpful for assessment of suitability of preformed aqueous foams to be used for other purposes, but they are not commonly used in the present field of application. The reason is that the behaviour of the foam in the

highly alkaline environment of the cement paste may differ
entirely from the behaviour under test conditions. Thus
e.g., a foam which had proved high stability when used for
fire fighting puposes, collapsed rather quickly when mixed
with cement paste. The only reliable way of measuring the
performance of a certain foam concentrate to be used for
the generation of preformed foam for making foamed concrete
is, therefore, testing the foam directly in the concrete.
Such a test method is described in detail in ASTM Deʋig-
nation C 796 - 87a. This test method includes the
following:

- Manufacture of laboratory quantities of a standard foamed
  concrete mix (cement content 400 kg/m^3, neat cement paste
  without using any aggregate, oven dry density about
  500 kg/m^3).

- Determination of the air content of the freshly prepared
  foamed concrete and of the hardened concrete after handling in conventional machinery (pumping through a 25-mm
  inside diameter open-end 15-m rubber hose, discharging
  the concrete in a sampling basin, the exit end of the
  hose being at the same height as the pump).

- Determination of significant properties of the hardened
  concrete, such as compressive strength, splitting tensile
  strength, density, and water absorption.

The foam concentrate is considered suitable for the production of foamed concrete, if the loss of air during pumping does not exceed 4.5 % by volume and if the requirements
of ASTM Designation C 869 - 80 with respect to strength and
water absorption of the hardened foamed concrete are ful-
filled.

Mix-foaming according to method (b) can be applied,
e.g., for transforming a mixture of lightweight aggregate
concrete with open structure into a no-fines lightweight
aggregate concrete mix with closed structure.
   Recently, a special air-entraining and plasticising admixture has been developed on the basis of proteins which
is capable of reducing the surface tension of the mixing
water and of introducing closed air pores into the cement
paste, thereby increasing the moveability, the water retention and the cohesion of the freshly mixed concrete. Other
than traditional air-entraining plasticizers, the new admixture does not increase the "yield". In contrary, by cre-
ating extremely dense package of the lightweight aggregate
particles, it reduces the overall volume of the mix, resulting in a higher strength/density ratio and improvement
of other important properties of the hardened concrete,
such as lower drying shrinkage.

(Note: "Yield is the relationship of the volume of concrete to loose volume of aggregate. A typical lightweight
aggregate concrete mix with open structure, designed for
1 m^3 of compacted concrete, may contain for example 1,05 m^3
loose expanded clay aggregate 4/8 mm, 0,2 m^3 lightweight
fines, 180 kg cement, and 90 to 110 kg water. On the other
hand, 1 m^3 of lightweight aggregate foamed concrete of the
type discussed in this contribution is typically composed
of 1,2 m^3 loose expanded clay 4/8 mm, 200 kg cement, and
100 to 145 kg water and is produced without adding of any
aggregate fines. The yield as defined above is only 0,83.
That means, that this type of lightweight aggregate foamed
concrete contains a 14 % higher volume of coarse lightweight aggregates per cubic metre than a traditional lightweight aggregate concrete mix with open structure.)

2.2 Mixing

2.2.1 Foamed concrete with preformed foam
For the production of foamed concrete with preformed foam
the following batching sequence is recommended:

At first the water is added to the running mixer followed by the cement and mixed until a homogeneous paste without lumps of undispersed cement has been obtained. Then any
sand and coarse lightweight aggregate is added and thoroughly mixed. If necessary, the water content is adjusted
in order to obtain a plastic consistency of the mixture.
Subsequently, while still mixing, the necessary volume V_f
of preformed foam is introduced. After all the foam has
been added, mixing should be continued for at least 2 min
in order to gain uniform distribution of the foam.

The required volume V_f of the foam to be added to obtain
a given density of the concrete may be assumed with good
approximation as the residual volume in the concrete which
is not occupied by the solid and liquid constituents (cement, aggregates, water, additions, if any). This is due to
the fact that, under favourable conditions, due to its high
stability, a quantity of aqueous air-foam containing a
definite volume of air and water, can be blended without
change in volume into the premixed slurry, mortar, or
lightweight aggregate concrete.

The water requirement will vary with the type and the
source of the cement, the mix proportions, and the grading
and water absorption of the aggregate. In general, it is
customary to gage the proper amount of water in a mix by
consistency rather than by a predetermined water-cement ratio. The water content, in combination with the added foam,
should be high enough to produce an easily pourable mixture. If the water content is too low, part of the air
cells will collapse during mixing, and the air-entraining
effect of the preformed foam will be reduced. In this case,
a higher quantity of foam will be required to achieve the

desired fresh density of the concrete mix. This will lead, more or less automatically, to an increase of the water content due to the fact that the foam consists by approximately 97 % of its mass of water forming the walls of the air cells. In general, when using preformed foam, it will therefore not be possible, to produce foamed concrete with other than pourable, nearly fluid consistency.

For small laboratory mixes the foam may be batched volumetrically, by means of calibrated vessels. In commercial applications, foam is injected from the foam generator directly into the mixer, and the quantity is gaged by duration of delivery. The necessary foam time is $V_f T_1$, where T_1 is the time required for the generation of 1 cubic metre of preformed foam, depending on the output of the foam generator which has been determined by calibration before.

Most types of mixers used for the production of ordinary concrete or mortar, such as pan mixers or rotating tilt-drum concrete mixers, are also suitable for mixing of foamed concrete. It is important, that the mixer is capable of causing enough vertical motion of the constituents in order to avoid that the light foam is acculumating in the upper area whereas the heavier constituents are remaining near the bottom. If pan mixers are used, their blades shall be set close to the bottom of the pan (not more than 3 mm away) to prevent that the cement paste adheres to the bottom. It is also essential that the mixer operator should be able to watch the batch being mixed so that he can judge water content, proper dispersion of the cement and uniform blending of the foam.

When transit mixing equipment is used for foamed concrete, the preformed foam should be added at the job site just prior to unloading, unless it is demonstrated that a mix of the required density and other properties can be delivered to the job site after adding the foam at the batching plant, ACI Committee 523 (1975). Experience has shown, however, that loss of air is generally small during transportation, even for long transportation periods, provided that a sufficiently stable foam has been used at the batching plant. Transit mixers need not be operated enroute to the job site. The loss of air during transportation and handling, if any, can be estimated in advance, e.g. by means of tests, and may be counteracted by increasing the quantity of foam to be added at the batching plant to such an extent that the design fresh density is obtained at the point of placing.

When the foam is added to the transit mixer at the job site, it should be blended with the premixed cement paste or mortar for at least 5 minutes at mixing speed. This is especially important when a water operated foam generator is used, which generally produces a foam with relatively larger pores. In this case, the size of the pores may be reduced by increasing the mixing period. It has been found that this may have a positive effect on the properties of the fresh and the hardened concrete.

2.2.2 Mix-foaming method
For the production of mix-foamed lightweight aggregate concrete tilt-drum mixers, such as installed in transit mixers, have been used with good success. With this mixer type, the risk is low that light and weak aggregate particles are getting crushed during mixing. On the other hand, these mixers are also capable of generating the necessary amount of entrained air within an acceptable period of time. Excessive mixing should be guarded against because of the possibility of changes in density, air content and consistency.

It is important that the mixer is kept free from impurities and from residues of previous mixes, e.g. set retarders, since such materials may disturb the generation of air pores or cause their collapse.

The recommended batching sequence is: dry premixing of the aggregates with the cement, then adding of the finely divided admixture and the water, and mixing until a pourable consistency has been reached. Experience has shown that a somewhat higher quantity of admixture may be required, if the admixture is added after the addition of the mixing water. As mentioned before, the proper quantity of admixture may have to be determined by trial and error for the particular mixer used and the concrete mix concerned. The resulting air content depends, among others, upon the type and gradation of aggregate, type and size of the mixer, size of batch, and duration of mixing. Allowance should be made for any decrease in air content between the end of mixing and placement of the concrete in the forms. This loss may vary with the type of the foam concentrate used.

With respect to the properties of this special kind of lightweight aggregate foamed concrete, it has been found essential to minimize the yield. This can be reached by limiting the air content in such a way that the volume of aerated cement paste is just sufficient to fill the interstitial space between the coarse aggregate particles. Any mayor surplus of aerated cement paste should be avoided.

3 Workability of freshly mixed foamed concretes

3.1 Workability properties
The workability of foamed concrete containing relatively large amounts of air pores is generally excellent. The mixes have usually an easily pourable, creamy or nearly fluid consistency. They are homogeneous, and, due to the high air content, bleeding and segregation problems will not ordinarily be present, Valore (1956). They will be

readily and easily placed simply by pouring and screeding, without the need of further consolidation. Because of their tendency of self-levelling, the application of those concretes may be limited to components with a fairly flat surface. Moulds or formwork should be reasonably watertight. More viscous mixtures, due to their good cohesion, are susceptible to entrapment of large air voids during moulding and may require slight compaction, therefore.

Also the foamed no-fines concretes as described in 2.1, where the interstitial voids between the aggregate particles are filled with a porous matrix containing closed air cells, show highly improved workability as compared with traditional lightweight aggregate concretes with open structure. The latter, because of their necessarily rather dry consistency, generally require special measures for adequate compaction, such as a vibrating table, sometimes combined with a roller device on the surface, whereas the foamed type has a pourable fluid consistency, a good cohesion, an improved water retention, and an excellent compactibility, which enables the concrete to be cast without additional compaction.

3.2 Measurement of workability and consistency

Appearance of the mixture may be a more reliable indication of consistency of foamed concrete than a test. In many cases, therefore, no tests are performed for the evaluation of the consistency. If testing is desired, the flow test according to ISO 9812 is a suitable test method. For foamed concrete without coarse aggregate it has been found advantageous to modify this test in that way that the spread is measured immediately after removal of the conical mould, without any shocking motion of the flow table. In this case, a spread of (600 to 700) mm has proved to be appropriate for most applications. If a sloped surface of the concrete is desired, the water content should be reduced, corresponding to a spread of (500 to 550) mm. The maximum slope which can be achieved is approximately (3 to 4) %. The tendency of self-levelling is lower in the very low density range, because of the reduced dead weight and the increased cohesion of the concrete as a consequence of the high air pore content.

ACI Committee 523 (1975) Guide for cellular concretes above 50 pcf, and for aggregate concretes above 50 pcf with compressive strengths less than 2500 psi. **ACI Journal** February 1975, 50-66.

Valore, R.C. JR. (1956) Insulating Concretes. **ACI Journal** November 1956, 509-532.

25 RHEOLOGY OF FRESH SHOTCRETE

D. BEAUPRÉ and S. MINDESS
Department of Civil Engineering, University
of British Columbia, Vancouver, BC, Canada

M. PIGEON
Département de Génie Civil, Université Laval, Québec, Canada

Abstract

This paper presents some theoretical considerations for studies on high performance wet-mix shotcrete. The existence of a flow resistance might be a good explanation for the "shootability" of shotcrete. An existing apparatus (MKIII) has been used as a basis for the design and construction of a new rheometer for determining the flow resistance. This new apparatus uses a data acquisition system to drive an impeller from rest to maximum speed, and then to rest again. Different speed increments can be used. The torque requirement and the angular velocity are continuously recorded. This procedure provides a better measurement than the usual straight line extrapolation method for determining the flow resistance.

Keywords: Shotcrete, Rheology, Rheometer, Yield stress, MKIII, Air content, Silica fume.

1 Introduction

Research on shotcrete has generally been carried out to find quick answers to specific problems. Very few studies have been in order to generate a fundamental understanding of the shooting process. As a result, we know how to apply thick coats of material efficiently, but there is still no clear answer as to why the shotcrete remains in place after shooting.

There are two possible answers to this question: the shootability may be explained by the compaction that occurs during the shooting process, or by the existence of a flow resistance in the shotcrete. It is most likely that both mechanisms act together.

Fresh concrete can be represented well as a Bingham material (Tattersall and Banfill, 1984). Figure 1 shows the usual representation of fresh concrete, which can be modelled by the equation:

$$T = g + h N,$$

where T is the torque required to drive an impeller at a certain angular velocity N in fresh concrete. The value g (which represents the flow resistance) is the intercept on the abscissa of a plot of N vs T, and h is the slope (which can be compared to "viscosity") of the line.

Special Concretes: Workability and Mixing. Edited by Peter J. M. Bartos. © RILEM.
Published by E & FN Spon, 2–6 Boundary Row, London SE1 8HN, 0 419 18870 3.

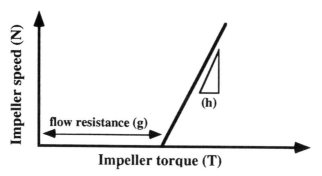

Figure 1: Fresh concrete behavior

Even if no clear answers have yet been demonstrated, it is reasonable to suppose that the existence of a flow resistance provides a good explanation as to why shotcrete is shootable. Intuitively, the higher the flow resistance, the better the shootability (i.e. the greater the thickness that can be built up without sloughing). In fact, a material with no flow resistance (such as water) could not remain in place after shooting. Similarly, a flowing concrete with a low flow resistance would not be suitable for shotcreting; it would simply slough off the receiving surface unless special agents (accelerators) were added at the nozzle. On the other hand, mixtures with a high flow resistance (low workability) could be unsuitable for shotcreting, because of pumping and consolidation difficulties.

The present work describes the development and use of a new rheometer to study the rheology of shotcrete, as a part of a research program to develop high performance shotcrete.

2 Development of high performance shotcrete

Recent developments in the production of high performance concrete (HPC) have pointed the way to the development of high performance wet-mix shotcrete (HPS). As is the case for HPC, the most logical way to make HPS is to use a superplasticizer (SP) to overcome the decrease in workability (increase in flow resistance) caused by the reduction of the water/cement ratio (W/C). The use of silica fume (SF) and good aggregates is also recommended. For instance, Kompen and Opsahl (1986) were successful in producing a high performance steel fibre reinforced shotcrete with a low W/C mix containing silica fume and superplasticizer. They obtained a compressive strength of 100 MPa measured on cubes.

The effect of W/C reduction, and the effects of SP and SF on variations in the flow resistance and viscosity of cement paste have been studied by Ivanov and Roshavelov (1990). A reduction in W/C and a decrease in SP dosage produced an increase in flow resistance and a reduction in viscosity. Silica fume at low dosages produced a modest decrease in the value of these parameters, but an intense increase at high dosages (between 7.5 to 15% cement replacement). Similar results have been found for concrete (Rixom and Waddicor, 1981; Banfill, 1980; Banfill, 1981; Tattersall, 1991).

It is possible to reduce the cement paste flow resistance to almost zero by using a high SP dosage (Banfill, 1980). Thus, in shotcrete, superplasticizers should be used with great care, and preferably with silica fume, to avoid shooting problems such as sloughing (or excessive set retardation).

In a series of tests on wet-mix asbestos fibre reinforced shotcrete, Beaupré et al. (1991) found that the best way to incorporate high volumes of asbestos fibres in the mixtures was to increase the paste volume by making mixtures with a high air content. Shotcretes with initial air contents as high as 20 % could be pumped and shot with very little rebound. The authors noted that the loss of air during pumping and shooting was very high, about 15% (i.e. leaving about 5 % air content in the hardened shotcrete). They also noted an almost complete absence of rebound (0.2% on walls and 0.5 % overhead) which indicates very good shootability. They concluded that, even if the presence of asbestos fibres in the mix has some value, the reduction in workability caused by the expulsion of air (also called compaction) is of greater interest. This loss of air during compaction caused a reduction in paste volume, which likely resulted in an increase in the flow resistance of the in-place shotcrete. Based on this information they formulated a concept by which it should be possible to make a pumpable shotcrete using a high initial air content, and still have good plastic and hardened properties for the in-place shotcrete because of the air losses during shooting. This concept involves surrounding the aggregates with a low W/C cement paste full of air bubbles. The bubbles have to be stable enough to resist pump pressure (to avoid blocking caused by pressure bleeding). On the other hand, the bubbles must not be too stable because they have to be broken down during compaction to prevent large reductions in the mechanical properties of the hardened shotcrete.

That study indicated that a high initial air content is uniquely suited to the wet-mix shotcrete process. High air contents in as-batched shotcrete apparently reduce the flow resistance, thus enhancing pumpability; on impacting on the receiving surface, air content is lost (increasing the flow resistance), which enhances shootability.

3 Developement of a new rheometer

Since the MKI apparatus developed by Tattersall, which consisted of a standard Hobart mixer with a power requirement measuring device, many rheometers for concrete have been built by different researchers. The second generation (called MKII) corrected the lack of precision in measuring the torque and also gave a wider range of speed settings (five instead of three). The MKIII apparatus (Tattersall, 1984) corrected the impeller motion of the MKII by adding a planetary motion. Wallevik and Gjorv (1990) have used tachometer, pressure transducer and a chart recorder to remove the operator influence in estimating the pressure (which gives the torque). Cabrera and Hopkins (1984) used strain gauges and a slip ring to measure the torque. They also used a chart recorder.

A new rheometer has been developed, based on the MKIII apparatus designed by Tattersall, which gives a planetary motion of the H-shaped impeller. Figure 2 shows a schematic diagram of the apparatus. A computer is used to drive a DC motor from rest to the desired speed and back to rest again. A tachometer placed between the motor and the reduction gear box (60 to 1) allows a reading of the speed of the impeller. The torque measuring device uses four strain gauges to measure the deflection of a small beam working in bending. The signal from the strain gauge bridge is amplified 470 times before a slip ring, located at the top of the shaft, transfers the signal from the rotating shaft to the data acquisition system.

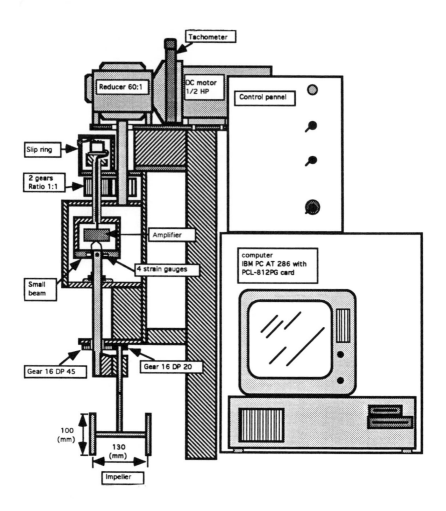

Figure 2: Schematic diagram of the new rheometer.

The computer program, which controls the test and the data acquisition board, allows one to customize the test parameters, such as the maximum speed setting, the speed increment, the speed decrement, the time for the motor to stabilize at a given speed, the sampling interval between speed and torque readings and finally the number of readings. To prevent damage to the rheometer, the test is automaticaly halted if a particular maximum torque is reached. A data file is created for each run, containing the motor voltage input, the impeller speed and the torque as well as the test parameters and a brief description of the mix.

Figure 3 gives an idea of the way in which a test is run. After taking a number (8 for this example) of readings of speed and torque, the program increases the voltage of the motor drive by an increment (here 300 in internal units). After a stabilization time (30

228

internal units which correspond to 1.7 sec), the computer executes another series of readings (speed and torque). Each reading is separated by a sampling interval (here 5 internal units or 0.3 sec). When the maximum speed chosen is reached, the speed is reduced step by step (here a decrement of 600) until zero speed is reached and the final value of torque is measured. At the end of the test, the computer stores all of the readings and the test parameters in a file for further analysis.

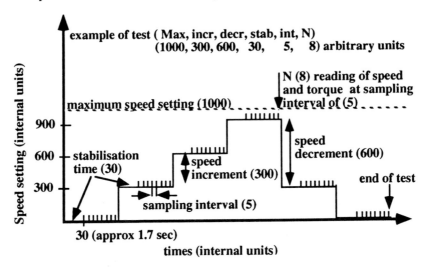

Figure 3: Schematic representation of test

4 Test Program

The first step of the project was to cast several concrete mixes to verify the behavior of the new aparatus and to determine the best parameters for testing. Also, this first step was intended to verify whether the extrapolations made using the "standard" MKII and MKIII rheometers are a valid means of determining the flow resistance.

The mix selection included normal plain shotcrete mixes and silica fume mixes with different air contents. Some additional mixes were made to check the effects of cement type combined with the use of small dosages of superplasticizer. The compositions of the various mixes are presented in Table 1. The mix identification gives the cement type (Type 10 (T10), Type 50 (T50) and Type 10 with replacement of 10% silica fume (SF)). This is followed by the W/C and a letter which indicates different the air contents. Because of some delays in the completion of the laboratory size concrete pump, only mix T10.40 has been shot (T10.40s), Nevertheless, other mixes especially T10.43 and SF.43 would probably have been excellent mixes for shooting purposes.

The results of some standard tests (slump, air content, specific weight, and 7-day and some 28-day compressive strengths measured on three 100 x 200 mm cylinders) as well as the determination of the flow resistance are given in Table 2. From those results, it is obvious that when the slump increases, the flow resistance is reduced. The flow resistance determination is discussed further in section 5 below.

Table 1: Mix Composition

mix	cement	Silica	water	aggregates*	W.R	A.E.A	SP
	(kg/m^3)	(kg/m^3)	(kg/m^3)	(kg/m^3)	(l/m^3)	(l/m^3)	(l/m^3)
T10.43a	412	0	176	1772	0.91	0	0
T10.43b	390	0	167	1679	0.86	0.20	0
T10.43c	380	0	162	1632	0.83	0.58	0
FS.43a	429	48	184	1590	1.15	0.80	0
FS.43b	359	40	154	1330	0.96	4.50	0
T10.38	463	0	177	1723	0	0	1.35
T50.38	464	0	174	1731	0	0	1.35
FS.38	411	46	170	1703	0	0	1.98
T10.40	351	0	140	1487**	1.25	2.30	0

* 50% sand and 50% 10 mm max size stone
** 60% sand and 40% 10 mm max size stone

Table 2: Test Results

mix	slump	air content	specific weight	strength (7 days)	strength (28 days)	flow resistance
	(mm)	(%)	(kg/m^3)	(MPa)	(MPa)	(Nm)
T10.43a	95	3.5	2364	49.3	67.3	2.6
T10.43b	240	8.0	2256	28.1	41.1	1.2
T10.43c	250	9.0	2202	29.8	39.7	1.0
FS.43a	60	5.5	2281	39.7	-	3.5
FS.43b	195	15.0	2023	18.0	-	1.4
T10.38	85	3.0	2395	48.4	-	3.8
T50.38	230	3.1	2392	46.0	-	1.3
FS.38	80	5.0	2364	47.2	-	3.6
T10.40	50	15.0	2114	19.5	-	2.2
T10.40s*	-	3.5	2378	30.7	-	3.2

*after being shot

5 Rheometer test results

Figures 4 to 12 show the results of several tests performed with the new rheometer. Except for Figure 6, each dot represents the average of all torque measurements at a fixed speed.

Figure 4 shows the results of a plain shotcrete mix. The results obtained as the speed is increasing (arrow pointing up on figs 4, 5, 7 and 8) are different from those obtained as the speed is decreasing. This behaviour is known as thixotropy.

The descending branch of the curve represents the kind of results obtained when using the Tattersall experimental method (Tattersall, 1991). In Figure 4, the descending branch, because of the limited number of measurements, is typical of the Tattersall results

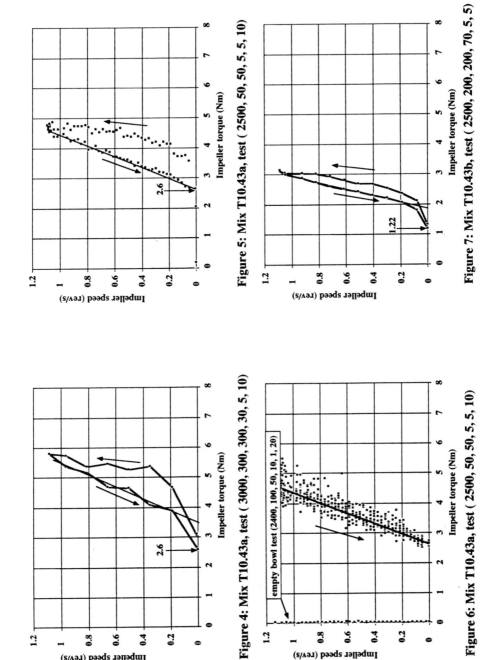

Figure 4: Mix T10.43a, test (3000, 300, 300, 30, 5, 10)

Figure 5: Mix T10.43a, test (2500, 50, 50, 5, 5, 10)

Figure 6: Mix T10.43a, test (2500, 50, 50, 5, 5, 10)

Figure 7: Mix T10.43b, test (2500, 200, 200, 70, 5, 5)

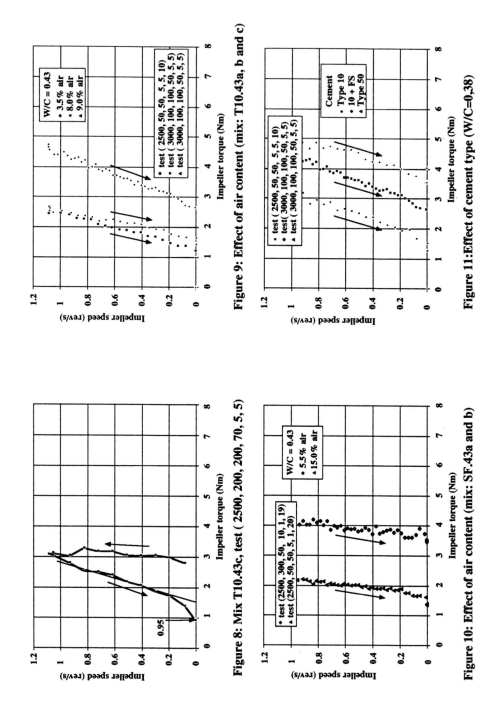

Figure 8: Mix T10.43c, test (2500, 200, 200, 70, 5, 5)

Figure 9: Effect of air content (mix: T10.43a, b and c)

Figure 10: Effect of air content (mix: SF.43a and b)

Figure 11:Effect of cement type (W/C=0,38)

if we remove the measurement at zero speed. Using the straight line extrapolation, the flow resistance of this concrete would be 3.5 Nm, but it is more likely 2.6 Nm (from the zero speed measurement). Figure 5 presents the results of a more precise test of the same mix (using a smaller speed increment and decrement as well as a greater number of readings at each speed). From this test, the flow resistance is again 2.6 Nm .

In order to visualize the spread of the data, all data from the descending part of the test shown in Figure 5 are plotted in Figure 6. These results are very close to those obtained by Cabrera and Hopkins (1984) using a chart recorder. An interesting point is that the amount of spread is reduced at lower speeds. A test without any concrete in the bowl was conducted to evaluate the effect of inertia and/or friction on the test results. The results (Fig. 6) indicate that these effects are very small (maximum reading of .07 Nm) and may be neglected.

Figures 7 and 8 show the results of tests on the same plain shotcrete mix but with higher air contents. On both curves, the extrapolation made without taking measurements at speeds lower than 0.2 rev/s overestimates the flow resistance.

In Figure 9, the effect of increasing the air content on the flow resistance is clearly shown. Figure 10 shows similar results for silica fume shotcrete mixes.

Figure 11 shows the effect of cement type and silica fume replacement. The use of Type 50 cement increases the workability by reducing the flow resistance. Its use in high performance shotcrete should thus be tested and should allow further reductions in the W/C. As expected, the replacement of cement by silica fume produces an increase in the flow resistance.

Figure 12 shows the effect of air loss (compaction) which occurs during shooting; this is primarily an increase in flow resistance without a significant change in viscosity when the air content drops from 15.0% to 3.5%. This reduction in air content is accompanied by an improvement compressive strength, from 19.5 MPa to 30.7 MPa, as shown in Table 2.

6 Conclusion

The new rheometer is very accurate in determining the flow resistance. The use of a high air content improves the workability by reducing the flow resistance. The compaction associated with the shooting process causes a reduction in air content which results in an increase in the flow resistance after shooting. The use of a high initial air content is a valid way to temporarily enhance the "pumpability" without impairing the final properties of the shotcrete. More studies are necessary to verify the applicability of this concept to high performance shotcrete.

7 Acknowledgement

This works has been funded by the Natural Science and Engineering Research Council of Canada, through the Networks of Centers of Excellence on High-Performance Concrete.

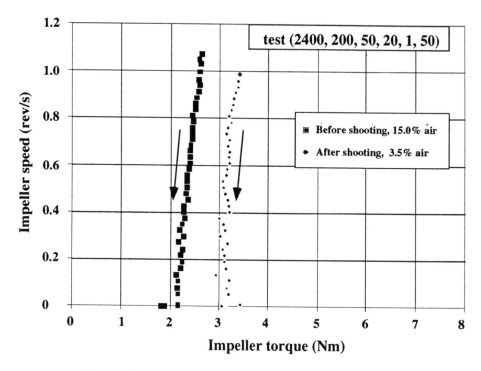

Figure 12: Mix T10.40 and before and after shooting (T10.40s)

8 References

Banfill, P.F.G. (1980) **Workability of flowing concrete,** Magazine of Concrete Research, 32 (110), 17-27.

Banfill, P.F.G. (1981) **A Viscosimetric Study of Cement Paste Containing Superplasticiser with a Note on Experimental Techniques,** Magazine of Concrete Research, 33 (114), 37-47.

Beaupré, D.; Pigeon, M.; Morgan, D.R. and McAskill, N. (1991) **Le Béton Projeté Renforcé de Fibres d'Amiante,** Proceedings of the First Canadian University-Industry Workshop on Fibre Reinforced Concrete, Université Laval, Québec, 28-29 Octobre, 1991.

Cabrera, J.G. and Hopkins, C.G. (1984) **A modification of the Tattersall two-point test apparatus for measuring concrete workability,** Magazine of Concrete Research, 36 (129), 337-40.

Ivanov, Y.P. and Roshavelov, T.T. (1990) **The Effect of Condensed Silica Fume on the Rheological Behavior of Cement Pastes,** Proceeding of the

International Conference organized by the British Society of Rheology, University of Liverpool, March 16-29, 1990.

Kompen, R. and Opsahl, O.A. (1986) **Wet-Process Shotcrete with Steel Fibre and Silica Fume - State of the Art in Norway**, 1986.

Rixom, M.R. and Waddicor, M.J. (1981) **The role of lignosulphonates as superplasticisers, in** Developments in Superplasticisers, Publication SP-68, Detroit, Americain Concrete Institute, 359-380.

Tattersall, G.H. and Banfill, P.F.G.(1983) **The Rheology of Fresh Concrete**, Pitman, London.

Tattersall, G. H. (1991) **Workability and Quality Control of Concrete**, Chapman & Hall, London.

Wallevik, O.H. and Gjorv, O.E. (1990) **Modification of the two-point workability apparatus**, Magazine of Concrete Research, 42 (152), 135-42.

PART EIGHT

MIXES FOR REPAIR OF CONCRETE

26 WORKABILITY AND MIXING OF MATERIALS FOR THE REPAIR OF DEFECTIVE CONCRETE

P. BENNISON
Flexcrete Limited, Preston, UK

Abstract
The workability and methods of mixing concrete repair materials varies considerably from semi dry materials to those which have to flow like a liquid.   The paper describes three categories for concrete repair mortars or concrete, those being: sprayed, hand-applied and flowing/pourable.   Each one of these categories requires its own special rheology and method of mixing and all these aspects are described in detail in the paper.
Keywords: Repair Mortar, Shotcrete, Gunite, Consistency, Mixing, Sprayed Concrete, Hand Applied Mortars.

1 Introduction

When considering methods for the repair of defective concrete, the first principle should be that the repair medium should be as close as possible in all physical and chemical characteristics to the concrete to be repaired.   The following must also be carefully examined:

Mechanical strength.
Protection of any steel reinforcement.
Adhesion between the repair and original concrete.
Bond between the repair and any steel reinforcement.
Shrinkage.
Thermal movement.
Durability.
Ease of application.
Cost.

Based upon the above criteria, concrete repair mortars or concrete fall into the general categories shown below:

Sprayed.
Hand-applied.
Flowing/pourable.

The workability parameters and related properties are therefore very different for materials in each of the above categories.   These range from very low water/cement ratio semi-dry mixes as used in the

Special Concretes: Workability and Mixing. Edited by Peter J. M. Bartos. © RILEM.
Published by E & FN Spon, 2–6 Boundary Row, London SE1 8HN, 0 419 18870 3.

dry process gunite, through sticky pseudoplastic or thixotropic hand-applied mortars, to free-flowing fluids for re-casting techniques.

2 Sprayed concrete and mortar

This method of application is particularly suitable for the reinstatement or repair of deteriorated or fire damaged concrete structures, where, by definition, large areas and volumes are involved, making other techniques uneconomic.

Definitions: 'Sprayed Concrete' is a mixture of cement, aggregate and water projected at high velocity from a nozzle into place to produce a dense homogeneous mass. 'Gunite' is a term used for sprayed concrete where the maximum aggregate size is less than 10mm. 'Shotcrete' is a term used for sprayed concrete where the maximum aggregate is 10mm or greater.

By nature of its method of placing, Gunite possesses properties not normally found in conventionally placed concrete.

Two distinct methods of spraying mortar and concrete exist, namely the dry mix and wet mix processes. The dry mix system is the original and still most commonly used. In this method a mixture of cement, sand and aggregate, if used, but with no added water, is fed into a special mechanical feeder and metered into a high pressure delivery hose. The high velocity air stream conveys the material to the nozzle which is equipped with a water injection system operated by a valve under the direct control of the nozzleman. The function of the nozzle is to convert the dry material into mixed mortar or concrete of the correct consistency. This consistency will depend upon the design of the dry powder components and the distance to impact from the nozzle, in general 0.5m to 1.2m, and the angle of placement; overhead, horizontal, etc.

In reality the dry powder materials can come in different forms, from cement and sand to totally pre-packaged mixes which may contain many additives, such as hydration accelerators, microsilica, fibres and polymers. Sands suitable for spraying (generally 5mm to fine medium grade) ideally contain 5-8% water, enough to guard against segregation in the hoses but not sufficient to block the machines. Typically, the sand, cement and aggregate and or additives are pre-mixed in a conventional concrete mixer or hand-mixed for smaller jobs. This pre-mix is then introduced into the spraying machine, of which there are several types.

In general these consist of a material feeding hopper with an air-driven agitator and a compressed air supply. Revolving rotor ports are filled with the gunite pre-mix by gravity from the machine hopper and discharged by air pressure into an air chamber and conveyed to the nozzle where water is added via a water ring.

An experienced nozzleman controls the water, minimizes re-bound of material during spraying and also slump in deep sections.

Fig.1.　Gunite spraying machine showing feed
hopper being filled with pre-mixed material

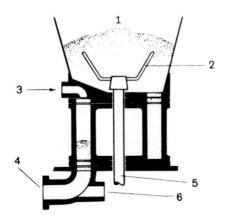

(1)　Material feeding/inlet
(2)　Agitator
(3)　Compressed air

(4)　Material exit
(5)　Drive shaft
(6)　Compressed air

Fig.2.　Diagram of the internal workings of a
Gunite spraying machine

Fig.3. The nozzleman spraying Gunite onto
an inclined concrete element

Dry process gunite as applied therefore has very low workability by
nature of low water/cement ratios (0.35-0.50) and method of
application.

In the wet process all the materials, including water, are
thoroughly pre-mixed prior to being introduced into the pumping
equipment.   Typically, simple forced-action pan mixers are used to
give a workability that allows the material to be sprayed along the
delivery hoses at relatively low velocity to the nozzle, which
converts the pumped material into a sprayed material by the
introduction of compressed air and not water as in the dry process.

Fig.4a. and Fig. 4b.   Creteangle forced-action pan mixer

3 Hand applied mortars

Strictly speaking, hand applied repair mortars is the expression used
when materials are literally packed into position with either a gloved

hand or trowel and finished with a steel or wooden float or polystyrene block.   This technique is applicable for medium/small repairs to, particularly, defective reinforced concrete and the filling of small surface defects (blow holes) prior to application of protective membranes.

As repairs have to be effected for differing depths and in the horizontal and vertical planes including overhead, the rheology and consistency of the materials vary considerably.   However, all the materials must comply with certain basic criteria as described below:

Durability:	The repair material must be at least as durable as the substrate concrete.
Protection of Steel:	Long term protection of any embedded metal is important for producing durable repairs.
Dimensional Stability:	Whilst curing the shrinkage must be minimal, thereafter any movement must be compatible with the substrate.
Ease of Application:	The durability of any repair will depend upon the ease of application, hence the number of rheologically different materials.

Historically, for this form of repair, individual materials were mixed on site to produce a repair mortar.   This originally was cement with a suitable sand and water mixed to a semi-dry consistency which would allow the material to be hand-compacted into the repair sites. However, because of this inherent lack of workability, only small areas and low application layer thicknesses could be achieved. Later, a polymer dispersion was substituted for plain water;   this introduction of tiny "plastic" film forming spheres, helped performance but still did not result in rates of application which were commercially acceptable.

However, when proprietary pre-packaged materials became available their technical sophistication allowed a much greater spectrum of durable repairs to be achieved.   Laboratory designed and quality assured factory manufactured materials allowed the use of additives which could modify and tailor the consistency and workability of the products to individual application and performance needs.

A variety of different polymer dispersions - Styrene Butadiene, Acrylic and Styrene Acrylic - are widely used to modify the internal cohesiveness in the plastic state and give other beneficial properties such as reduction in porosity and permeability in the hardened state.

Long chain, high polymers, such as hydroxy ethyl cellulose or hydroxy methyl cellulose can induce a degree of thixotropy and, when used in conjunction with a variety of different fibres, produce material with very low "sag" characteristics.   As many products incorporate the use of pozzolans, such as microsilica (a very fine material of mean particle size $0.1\mu m$), powerful de-flocculants are needed to make efficient use and gain maximum benefit.   These, in fact, are in general the widely used workability modifiers, such as sulphonated melamine formaldehyde condensates and sulphonated naphthalene formaldehyde condensates.

Air is sometimes purposely entrained to reduce density, although a more common way of reducing density to produce lightweight materials for overhead applications is by the incorporation of lightweight fillers.

Whilst in the "site-batched" materials, mixing is a very hit and miss affair (little measuring or weighing), relying upon the operatives experience to produce a consistent mortar, which is usually mixed by hand on a board. Many pre-packaged mortars come in the form of two components, a bottle of polymer dispersion and a bag of powder, others may be single component, requiring the addition of clean water. In order to mobilise the constituents properly, neither system can be satisfactorily hand-mixed, requiring the use of higher energy, usually with either a slow speed drill and paddle or a forced action pan mixer.

For hand packing techniques where the repair materials are compacted into medium sized voids, the materials used tend to be very cohesive or sticky, many incorporating a 10mm coarse aggregate.

However, it is important that the surface of the repair can be finished with a steel float or similar without inducing "dragging" in order to obtain an acceptable surface texture to the repair. When temporary support and shuttering is not used, the material should have a consistency such that it does not sag when applied in commercially acceptable application layers. Pre-packaged materials are available

Fig.5a. and 5b. Slow speed drill and paddle for
mixing repair materials

Fig.6. Different designs of mixing paddles

which can fulfil all the required parameters and be applied in single
layers up to 100mm thick even in soffit situations.

**Fig.7. Filling and compacting of a beam repair
using the hand-packing technique**

Fig.8. **Hand-applied repair being carried out**
without any form of shuttering

Direct application of mortar with a steel float or bull-nose trowel
is commonplace for small repairs where shuttering is not used.   In
this instance the mortar must have properties which allow it to be
cohesive enough to be placed in high application layers (up to 80mm)
without slumping or disbonding (even in overhead applications), yet
not so "sticky" that it "grabs" to the steel float surface.

Small surface imperfections, such as blow holes in the concrete,
require filling prior to the application of surface coatings, if this
is not carried out then pressure may cause a bubble and subsequent
pin-hole in the coating.   The most common technique for filling these
defects is known as "bag rubbing".   Stiff, yet cohesive mortar is
literally rubbed into the surface of the concrete using a piece of
hessian, stiff cloth or polystyrene block.

3 Flowing/pourable mortars

When using re-casting techniques for the reinstatement of damaged,
defective or deteriorated concrete by partial or total replacement,
high strength, non-shrink, flowing "concrete" is usually employed.
This method is employed for large scale reinstatement where the
surface profile of the finished repair must be aesthetically
acceptable, such as highway bridge structures, although some smaller
repairs may be suited to this technique.   However, when partial
replacement is being undertaken, this method has potential inherent
disadvantages.

The consistency of the mix must allow it to be either pumped into
place or poured under gravity where it must self-compact to eliminate
surface defects, such as honeycombing, and then effect a good bond to
the prepared concrete substrate.   The material used is usually a high
strength grout, which employs a dual expansion system which
compensates for shrinkage in both the plastic and hardened states, and
non-reactive 5-10mm aggregate.

Fig.9. Medium sized repair suitable for the use
of "flowing concrete" type material prior to the fixing of
the side shutters and the letter box placement openings

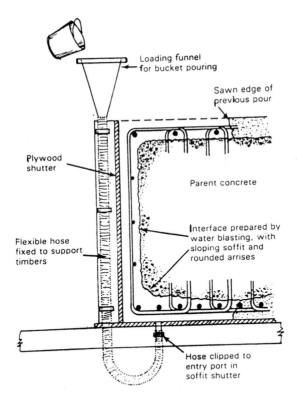

Loading funnel
for bucket pouring

Sawn edge of
previous pour

Plywood
shutter

Parent concrete

Interface prepared by
water blasting, with
sloping soffit and
rounded arrises

Flexible hose
fixed to support
timbers

Hose clipped to
entry port in
soffit shutter

Fig.10. Diagram of a poured under pressure placement
of a high performance, non-shrink, flowing concrete

If the material to be used is a "traditional" concrete, then extreme care must be taken with the mix design, the high mobility of concrete should be achieved by the use of high grade water-reducing admixtures and not by a high water/cement ratio. Placing should be in accordance with good working practice with particular attention being paid to compaction.

The proprietary materials are mixed using methods described earlier - slow speed drill and paddle or forced action pan mixers - to give a consistency similar to "pea soup". Traditional high mobility concrete is most commonly placed using either "letter box" or "fixed pouch" type methods. High performance, non-shrink, flowing concrete, is usually placed using a poured under pressure technique, but gravity placement by "letter box" is equally applicable.

4 Conclusions

With the growth of the necessity to repair defective, reinforced concrete structures, there has come a much greater interest in the workability parameters and related properties of the materials used for such repairs. This paper has described three basic approaches which require materials with application properties very different from traditional concrete, but which must fulfil the role of a homogeneous concrete substitute for the life of the structure.

5 References

Association of Gunite Contractors.
Aliva Ltd., Hatfield.
Bennison, P. (1987) Repair and Protection of Reinforced Concrete Bridges, in **Concrete Bridge Engineering: Performance & Advances**, Elsevier Science Publishers Ltd, pp. 107-141.
Cement and Concrete Association, **Current Practice Sheets 3PC/06/1**.
Palladian Publications Ltd., **Concrete Repairs Volume 3 (1989)**.

27 EVALUATING THE FLOW PROPERTIES OF FLOWABLE CONCRETE

A. McLEISH
WS Atkins Consultants, Epsom, UK

Abstract
The approach adopted for the assessment of flow properties of flowable repair concrete for use in the reinstatement of a corrosion damaged viaduct is discussed. The use of initial compliance tests to ensure that the concrete meets the specifications, production control tests on each batch of concrete and routine site tests is outlined.

1 Introduction

This paper discusses the approach adopted for ensuring that the properties of a flowable concrete were appropriate for repairs to a series of corrosion damaged reinforced concrete viaducts.

Although the paper concentrates on the flow properties it should be remembered that repair concrete often demands properties not generally considered in detail in concrete for new works and in particular it must:

Have a small maximum aggregate size to allow flow in the restricted spaces between parent concrete, reinforcement and formwork;
Be self levelling and self compacting as access for vibrators is often limited;
Have a high rate of strength gain, particularly where a lot of small localised repairs have to be undertaken;
Be dense and have a low permeability as often repairs result from a combination of low cover and chloride/carbonation attack of the original construction which must be overcome;
Not suffer from any bleed that would destroy the bond of the repair concrete to the parent concrete, thus reducing the structure effectiveness and presenting a leakage path into the reinforcement.

Special Concretes: Workability and Mixing. Edited by Peter J. M. Bartos. © RILEM.
Published by E & FN Spon, 2–6 Boundary Row, London SE1 8HN, 0 419 18870 3.

2 General requirements for repair concrete

A problem had arisen on a motorway flyover because of
deteriorated joints which had allowed the penetration of
deicing salts from the road deck above onto the supporting
reinforced concrete beams.

To avoid extensive and complex temporary support, the
repairs had to be carried out piecemeal such that only
small areas could be repaired at any one time. This
necessitated concrete with a high rate of strength gain to
minimise delays between successive repairs.

Access difficulties precluded the use of poured
concrete and resulted in the choice of a flowable, self
levelling, self compacting concrete. Congested
reinforcement and low cover in some areas dictated a
maximum aggregate size of 8mm.

The key elements of the specification for the flowable
repair concrete were:

Minimum cement content	450 Kg/m³
Maximum aggregate size	8mm
Maximum water/cement ratio	0.40
Minimum compressive strength	30 N/mm² at 3 days at 20°C
Maximum compressive strength	60 N/mm² at 7 days at 20°C

To ensure good quality control it was decided to use
proprietary factory dry batched repair concrete. Details
of the various repair concrete mixes used remain
confidential to the manufacturers. However in general the
mix constituents were as follows:

OPC in range 480 to 550 kg/m³
20% Pfa or 30% GGBS or 5% microsilica
8mm aggregate in range 1220 to 1320 kg/m³
W/C ratio around 0.35 to 0.40
Plasticiser and other undefined admixtures

3 Testing philosophy

Early in the development of the specification for the
repairs it was realised that the required properties of
the flowable concrete were highly sensitive to small
changes in the constituents and in any case difficult to
achieve. A three stage testing regime was therefore
adopted to ensure compliance:

Initial Compliance Tests	These consisted of extensive laboratory and field trials of the concrete to ensure full compliance with the specification

| Production Control Tests | Each batch of concrete manufactured was tested for flow and compressive strength |
| Site Tests | Routine site tests on flow and strength to compare with previous initial compliance and production control test results |

3.1 Initial Compliance Tests
3.1.1 Flow Trough Tests
The flow characteristics were assessed using the flow trough equipment shown in Figure 1. This consists of a horizontal steel trough 230mm in width and 1000mm long. A conical hopper is charged with 6 litres of concrete and a plug released to allow the concrete to fall into the trough and flow along its length. Each test consisted of six readings, three taken immediately after completion of mixing and three taken 30 minutes later. For compliance none of the flow test times was to exceed 30 seconds for flow along a distance of 750mm (at 20° C).

Figure 1 - Flow trough test

To evaluate the effects of temperature and delay between mixing and carrying out the flow test, preliminary trials were carried out on a range of proprietary mixes. The results for tests carried out immediately after mixing are summarised in Table 1. Figure 2 shows the effect of time on the flow test results.

Table 1 - Flow test results

Material	Water (1/25 kg)	Temperature (°C)	Flow Trough	Simulated Soffit Test
A	3.5	5	690mm in 30 secs	Some trapped air pockets
	3.5	12	750mm in 12 secs	Fewer air pockets
	3.5	20	750mm in 5 secs	Fewer air pockets
B	2.2	5	750mm in 30 secs	Many air pockets
	2.2	12	750mm in 15 secs	Many air pockets
	2.2	20	750mm in 10 secs	Many air pockets
C	3.33	5	825mm in 30 secs	Some large air pockets
	3.33	12	1000mm in 17 secs	Some large air pockets
	3.33	20	1000mm in 8 secs	No air but white powder on top
D	2.25	5	858mm in 30 secs	A few air pockets
	2.25	12	795mm in 30 secs	A few air pockets
	2.20	20	798mm in 30 secs	More air pockets
E	2.5	5	743mm in 30 secs	Two cracks, no air pockets
	2.375	12	720mm in 30 secs	Many air pockets
	2.375	20	767mm in 30 secs	A few large air pockets
F	3.75	5	655mm in 30 secs	Satisfactory
	3.65	12	716mm in 30 secs	Satisfactory
	3.65	20	738mm in 30 secs	Many small air pockets

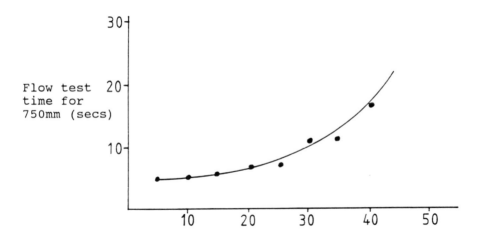

Time after water was added to mix (minutes)

Figure 2 - Effect of delay on flow trough results

It was apparent from these tests that whilst adequate flow could be achieved, the overall performance of the mix was very dependent on the precise water content of the mix and the temperature during placement. It was found that the effect of reducing the temperature from 20° to 5°C could reduce the flow properties by anything between 20% and 80% depending on the mix. Delays of up to 30 minutes between mixing and placing of many of the mixes were found to have little effect on the flow properties. Increased delays were however found to reduce the flow substantially.

3.1.2 Simulated Soffit Tests
The flow characteristics were also assessed by simulated soffit tests in which the flow of the concrete around reinforcement was checked through a perspex sheet representing the broken back soffit of the beam being repaired. The general arrangement of this test is shown in Figure 3. The layout of the reinforcement for this test was selected to be the most onerous in terms of achieving placement of the concrete that was likely to be encountered.

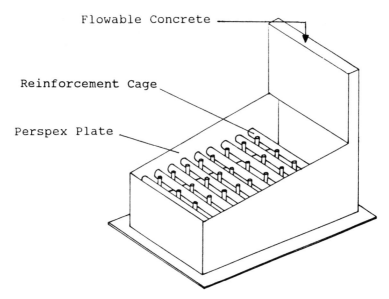

Flowable Concrete

Reinforcement Cage

Perspex Plate

Figure 3 - Simulated soffit tests

After the concrete had set, selected specimens were saw-cut into two sections which were examined to assess the amount of voidage around the reinforcement and at the repair/substrate interface, bleed at the interface, cracks and any other defects. Figure 4 shows the saw cut side and the top surface of the repair. Compaction of the concrete around the reinforcement was shown to be satisfactory although in this example numerous air pockets had formed on the top face, ie. at what would have been the interface between the repair and parent concretes.

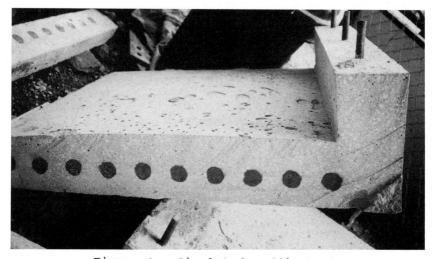

Figure 4 - Simulated soffit test

3.1.3 Preliminary Field Trials

To provide greater assurance that when used in a realistic repair situation the flowable concrete would perform satisfactory, a series of full scale repair trials were carried out. Full scale reinforced concrete beams were cast to enable the complete repair procedure to be evaluated. (See Figures 5 and 6).

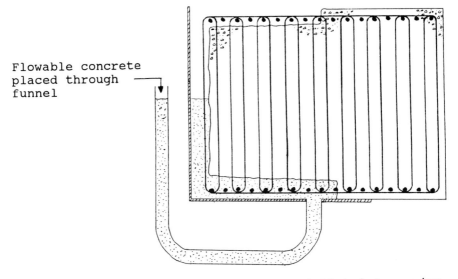

Flowable concrete placed through funnel

Figure 5 - General arrangement for full trial repairs

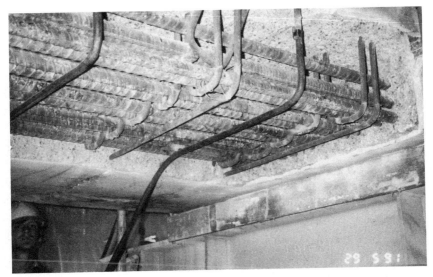

Figure 6 - Soffit reinforcement

These were then repaired by placing the flowable concrete through a flexible tube and funnel (Figure 7). On completion of the repair, cores were drilled horizontally in at the interface between the broken-back soffit and the repair concrete. These confirmed that a good bond had been achieved with minimal bleed or air bubbles (Figure 8).

Figure 7 - Placement by funnel and tube

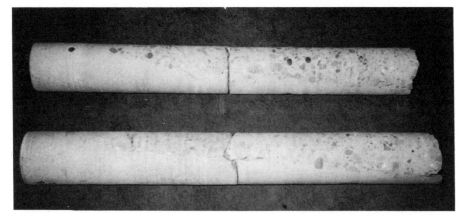

Figure 8 - Cores taken into trial repair

To investigate the manner in which the repair concrete
flowed within the formwork and around reinforcement trials
were carried out using coloured concretes. Figure 10
shows the results of one trial which indicated that
initial mixes do not move significantly after they have
flowed into the formwork, rather, the subsequent mixes
flow through them in a piping action.

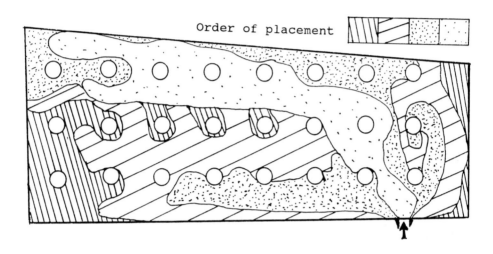

Figure 9
Section of simulated soffit pour using coloured dyes

3.2 Production Control Tests

Following the initial compliance tests various concrete
formulations were accepted as being suitable for use in
repairs to the structure. During the course of the repair
works production control tests were carried out on each
new batch of concrete supplied by the manufacturer.
Production control tests were restricted to flow trough
and compressive strength testing, both carried out at 20°C
only. These tests proved to be a very valuable method of
detecting poor quality or defective batches of material
and providing a check on the manufacturer's quality
control.

3.3 Site Tests

Whilst the production control tests demonstrated the
suitability of each batch of repair concrete further site
tests were carried out on site on every pour.

For each pour of concrete a flow test was carried out
and the compressive strength gain was monitored by testing
cubes stored alongside the repair areas at ambient
temperature.

AUTHOR INDEX

SUBJECT INDEX

This index has been compiled from the keywords assigned to the papers by the authors, edited and extended as appropriate. The numbers refer to the first page number of the relevant paper.

RILEM, The International Union of Testing and Research Laboratories for Materials and Structures, is an international, non-governmental technical association whose vocation is to contribute to progress in the construction sciences, techniques and industries, essentially by means of the communication it fosters between research and practice. RILEM activity therefore aims at developing the knowledge of properties of materials and performance of structures, at defining the means for their assessment in laboratory and service conditions and at unifying measurement and testing methods used with this objective.

RILEM was founded in 1947, and has a membership of over 900 in some 80 countries. It forms an institutional framework for cooperation by experts to:

- optimise and harmonise test methods for measuring properties and performance of building and civil engineering materials and structures under laboratory and service environments;
- prepare technical recommendations for testing methods;
- prepare state-of-the-art reports to identify further research needs.

RILEM members include the leading building research and testing laboratories around the world, industrial research, manufacturing and contracting interests as well as a significant number of individual members, from industry and universities. RILEM's focus is on construction materials and their use in buildings and civil engineering structures, covering all phases of the building process from manufacture to use and recycling of materials.

RILEM meets these objectives though the work of its technical committees. Symposia, workshops and seminars are organised to facilitate the exchange of information and dissemination of knowledge. RILEM's primary output are technical recommendations. RILEM also publishes the journal *Materials and Structures* which provides a further avenue for reporting the work of its committees. Details are given below. Many other publications, in the form of reports, monographs, symposia and workshop proceedings, are produced.

Details of RILEM membership may be obtained from RILEM, École Normale Supérieure, Pavillon du Crous, 61, avenue du Pdt Wilson, 94235 Cachan Cedex, France.

RILEM Reports, Proceedings and other publications are listed below. Full details may be obtained from E & F N Spon, 2-6 Boundary Row, London SE1 8HN, Tel: (0)71-865 0066, Fax: (0)71-522 9623.

Materials and Structures

RILEM's journal, *Materials and Structures*, is published by E & F N Spon on behalf of RILEM. The journal was founded in 1968, and is a leading journal of record for current research in the properties and performance of building materials and structures, standardization of test methods, and the application of research results to the structural use of materials in building and civil engineering applications.

The papers are selected by an international Editorial Committee to conform with the highest research standards. As well as submitted papers from research and industry, the Journal publishes Reports and Recommendations prepared buy RILEM Technical Committees, together with news of other RILEM activities.

Materials and Structures is published ten times a year (ISSN 0025-5432) and sample copy requests and subscription enquiries should be sent to: E & F N Spon, 2-6 Boundary Row, London SE1 8HN, Tel: (0)71-865 0066, Fax: (0)71-522 9623; or Journals Promotion Department, Chapman & Hall Inc, One Penn Plaza, 41st Floor, New York, NY 10119, USA, Tel: (212) 564 1060, Fax: (212) 564 1505.

RILEM Reports

1. Soiling and Cleaning of Building Facades
2. Corrosion of Steel in Concrete
3. Fracture Mechanics of Concrete Structures - From Theory to Applications
4. Geomembranes - Identification and Performance Testing
5. Fracture Mechanics Test Methods for Concrete
6. Recycling of Demolished Concrete and Masonry
7. Fly Ash in Concrete - Properties and Performance
8. Creep in Timber Structures
9. Disaster Planning, Structural Assessment, Demolition and Recycling
10. Application of Admixtures in Concrete
11. Interfaces in Cementitious Composites

RILEM Proceedings

1. Adhesion between Polymers and Concrete.
2. From Materials Science to Construction Materials Engineering
3. Durability of Geotextiles
4. Demolition and Reuse of Concrete and Masonry
5. Admixtures for Concrete - Improvement of Properties
6. Analysis of Concrete Structures by Fracture Mechanics
7. Vegetable Plants and their Fibres as Building Materials
8. Mechanical Tests for Bituminous Mixes
9. Test Quality for Construction, Materials and Structures
10. Properties of Fresh Concrete
11. Testing during Concrete Construction
12. Testing of Metals for Structures
13. Fracture Processes in Concrete, Rock and Ceramics
14. Quality Control of Concrete Structures
15. High Performance Fiber Reinforced Cement Composites
16. Hydration and Setting of Cements
17. Fibre Reinforced Cement and Concrete
18. Interfaces in Cementitious Composites
19. Concrete in Hot Climates
20. Reflective Cracking in Pavements - State of the Art and Design Recommendations
21. Conservation of Stone and other Materials
22. Creep and Shrinkage of Concrete
23. Demolition and Reuse of Concrete and Masonry
24. Special Concretes - Workability and Mixing

RILEM Recommendations and Recommended Practice

RILEM Technical Recommendations for the Testing and Use of Construction Materials

Autoclaved Aerated Concrete - Properties, Testing and Design

Workability and Quality Control of Concrete

G H Tattersall

Department of Civil and Structural Engineering, University of Sheffield, UK

This book is a successor to the author's highly successful 'Workability of Concrete' published in 1976 and, again, is written specifically for practising engineers and concrete technologists. It incorporates the results of a further 15 years research and develops the treatment of workability as a property to be measured in terms of two constants, which was introduced in the earlier book. The scientific basis is simply explained and used for the description of very practical methods and apparatus. This leads to elucidation of problems surrounding the topic of workability and to an account of the potential for quality control. The validity and limitations of standard methods of workability assessment are fully considered and there are several chapters on the effects on workability of the properties and proportions of mix constituents.

Contents: Preface. The importance of workability. Standard tests for workability. Flow properties of fresh concrete. Principles of measurement. The two-point workability test. Workability expressed in terms of two constants. Extremely low workability concretes. Factors affecting workability. Effect of mix proportions. Effect of chemical admixtures. Effect of cement replacements and fibres. Workability and practical processes. Specification of workability. Workability measurement as a means of quality control. Epilogue. Glossary. Index.

October 1991: 234x156: 272pp, approx. 110 line diagrams and 12 photographs
Hardback: 0-419-14860-4: £29.00

E & F N Spon
An imprint of Chapman & Hall

Properties of Fresh Concrete

Proceedings of the International RILEM Colloquium

Edited by **Professor H J Wierig**, Institute of Building Materials and Materials Testing, University of Hanover, Germany

The production of concrete has in recent years shifted from the site to central production plants. This has improved efficiency but has also given rise to new problems, such as the influence of temperature and time on consistency of fresh concrete. The separation of mixing and transportation from on-site placing and curing requires better means of measuring and defining the state of the concrete at the time of delivery. In addition, new types of cements and admixtures are being widely used.

This book forms the Proceedings of the RILEM Colloquium held in Hanover, Germany in October 1990 to review the state-of-the-art of the properties of fresh concrete. Papers from 18 countries in Europe, North America and the Far East are included.

Contents: Preface. **Part 1**: General properties of fresh concrete, mortar and cement. **Part 2**: Factors influencing the properties of fresh concrete. **Part 3**: Test methods: significance and reliability. Part 4: Interactions between properties of fresh concrete and hardened concrete. Part 5: Computer aided mix design and production. Index.

RILEM Proceedings 10

October 1990: 234x156: 400pp, 183 line diagrams, 21 photographs
Hardback: 0-412-37430-7: £45.00

Chapman & Hall

Rheology of Fresh Cement and Concrete

Proceedings of an International Conference, Liverpool, 1990

Edited by **P F G Banfill**, School of Architecture and Building Engineering, University of Liverpool, UK

Understanding and controlling the rheology of cement-based materials is essential if they are to be used successfully in building, civil engineering and offshore applications. This book brings together new research information on the flow behaviour of cementitious materials from the UK, France, Italy, Germany, Poland, Finland, USSR, USA and Japan, presented at the International Conference organised by the British Society of Rheology in March 1990. Topics covered include: measurement techniques and fundamental studies of rheology; hydration, setting kinetics and computer simulation; materials (oilwell cements, grouts, mortars, flyash); practical applications (vibration, early-age properties, formwork pressures).

Contents: Preface. **Part 1**: Cement pastes: effects of chemical composition, admixtures and latent hydraulic binders. **Part 2**: Theoretical studies. **Part 3**: Properties of oilwell cement slurries and cementing processes. **Part 4**: Properties of grouts and grouting processes. **Part 5**: Concretes. **Part 6**: Influence of vibration on cement-based systems. **Part 7**: Discussion. Index.

September 1990: 234x156: 384pp, numerous line diagrams
Hardback: 0-419-15360-8: £47.00

E & F N Spon
An imprint of Chapman & Hall

Structural Lightweight Aggregate Concrete

Edited by **J L Clarke,** Senior Engineer, Special Structures Dept., Sir William Halcrow & Partners, London, UK, formerly of British Cement Assocation, UK

Lightweight aggregate concrete is undergoing something of a renaissance. Although this material has been available for many years, only now is it being used more widely. The volume of structural aggregate concrete used each year is increasing dramatically. Lower structural weight, better fire resistence, use of waste for aggregate, lower costs for aggregate: all these factors are contributing to the rapid increase in the use of acceptance of structural lightweight aggregate concrete. This book provides a comprehensive review of this growing field from an international perspective.

"This volume provides a comprehensive review of the subject from an international perspective." - *British Bookseller*

July 1993: 234x156: 256pp, 80 line illus, 20 halftone illus
Hardback: 0-7514-0006-8

Blackie Academic & Professional
An imprint of Chapman & Hall

Testing During Concrete Construction

Proceedings of RILEM Colloquium, Darmstadt, March 1990

Edited by **Professor H W Reinhardt**, Building Materials Institute,University of Stuttgart and Managing Director of the Otto Graf Institute (Research and Testing Institute) of the State of Baden-Wurttemberg, Germany

Every structure has its requirements with respect to strength, appearance, durability, serviceability and to meet these requirements appropriate steps must be taken at all the stages of design, analysis, construction, inspection and maintenance. The properties of concrete are often related to the strength of test specimens at 28 days. However, strength tests give no information about the real concrete in the structure, such as the materials used, the mix design, compaction or curing. Yet the surface of concrete has a crucial influence on a structure's performance. Testing during construction has clear benefits:

- the quality of the work can be checked at a time when the building process can still be changed in order to improve the end product.

- the effect of changes in construction processes can be assessed.

- decisions can be taken on the basis of valid information.

Numerous disputes arise from the fact that the materials and mix proportions are not known, and from doubts about the cover to reinforcing steel. Testing of concrete during construction helps to assess and improve the quality of structures, to reduce overall costs, and to demonstrate the industry's concern to achieve high quality structures. This book covers testing and measuring techniques for fresh concrete, early-age concrete, reinforcement and prestressing tendons, especially non-destructive techniques. Accuracy, sensitivity, reliability and ease of use of the test methods are discussed and backed up by practical examples. Many examples are given of practical methods which can be applied directly or which suggest directions of future developments.

Contents: Preface. Part 1: Introduction. Part 2: Constituents in the mix. Part 3: Setting and compaction. Part 4: Early age strength development. Part 5: Durability-related testing. Part 6: Tests related to reinforcement and prestressing. Part 7: Evaluation of measurements. Part 8: Conclusion. Index.

RILEM Proceedings 11

January 1991: 234x156: 480pages, c.100 illus
Hardback: 0-412-39270-4: £49.00

Chapman & Hall